SEW ELECTRIC

A COLLECTION OF DIY PROJECTS THAT COMBINE
FABRIC, ELECTRONICS, AND PROGRAMMING

Sew Electric

a Collection of DIY Projects that Combine Fabric, Electronics, and Sewing
by Leah Buechley and Kanjun Qiu
Illustrated and designed by Sonja de Boer

This publication uses the font PlainPrint, which was created by and licensed from Katherine Reynolds. See: http://thirdgradeteacherfiles.blogspot.com.

Published by HLT Press, Cambridge, MA 02139

Sew Electric team
Designer and Illustrator: Sonja de Boer
Copy Editor and Indexer: Paula Szocik
Publishing Coordinator: Martha Baum
Technical and Editorial Assistant: Tayo Falase

Printed in the United States of America

ISBN-10: 0989795608
ISBN-13: 978-0-9897956-0-9

CONTENTS

FOREWORD, Jocelyn Goldfein   V

ACKNOWLEDGMENTS   VII

INTRODUCTION   1

- -

PROJECTS

BOOKMARK BOOK LIGHT   5

SPARKLE BRACELET   25

PROGRAMMING YOUR LILYPAD   41

INTERACTIVE STUFFED MONSTER   63

FABRIC PIANO   107

- -

TROUBLESHOOTING SOLUTIONS

ELECTRICAL PROBLEMS   151

CODE PROBLEMS, COMPILE ERRORS   155

CODE PROBLEMS, LOGICAL ERRORS   160

- -

MATERIALS REFERENCE, GLOSSARY, & INDEX

MATERIALS REFERENCE, electronic materials & tools   168

MATERIALS REFERENCE, craft materials & tools   170

GLOSSARY   172

INDEX   179

IV

It never occurred to me that I could learn how to make software. I was a pretty nerdy kid who loved science fiction and video games, but programming had this mystique. It was for prodigies in their parents' basements hacking into the CIA...not for kids like me.

I didn't learn to program until college, and at first it seemed like just another academic hurdle. I enjoyed my computer science classes, but mostly out of my love of logic and puzzle solving. This sounds obvious in hindsight, but it was only after my first software internship that I understood that *programming is about making things*. Programming is about designing and building things. It's about creative expression.

Working at Netscape at the beginning of the Internet revolution was even more transformative for me. I realized that the code I was writing wasn't just a homework assignment. I was making real things (a web browser!) that would be used by real people (my family! my friends! strangers around the globe!) My designs were limited only by my imagination. I could be playful or understated in my creative choices. I could open up the web to entire new countries if I tinkered with a few lines of code. I had found my calling.

Some of my classmates weren't as lucky as I was. They didn't have the chance to explore programming until much later in their lives, if ever. Technology seemed "too hard" or "not for them". The stereotypes around technology discouraged people who didn't want to be identified as tech geeks. Many people still imagine that technology is inaccessible and difficult to learn. Women in particular often find the "hacker" image of the stereotypical programmer off-putting.

Enter electronic textiles. E-textiles' blend of craft and technology appeals to all sorts of people—to people who consider themselves creators, designers, and artists as well as to "geeks" and "nerds". Fundamentally, craft and technology design are the same kind of activity. They are both fueled by the basic human urge to create. Is the growing e-textile community sewing, or are they laying down circuits? Designing for utility, or beauty? Is their work craft, or is it technology? The answer is YES. They are, very simply, making.

This book teaches you how to build electronics and program through hands-on e-textile projects. It's friendly and fun, introducing concepts as you need them. You'll want to figure out how to make your stuffed monster sing songs, and how to make your piano play different notes, and, by the end, you'll have learned to code without even realizing it. If you were a programmer to begin with, this collection of projects will introduce you to surprising new materials, tools, and techniques.

This is a book I would have loved when I was younger, and it would have inspired me to love electronics and programming much earlier in my life. So I encourage you to explore, and experiment—build the projects in this book, and modify them to your heart's content. Then, go beyond the book and think about what else you can make with the LilyPad Arduino and e-textiles: Add them to your clothing or furniture and combine them with other stuff. With the skills that you'll develop, the possibilities for what you can create are endless. You'll learn a ton about electronics and programming and build some cool stuff, but you'll also realize that you, too, can design and create the new technologies of the future. Enjoy!

- Jocelyn Goldfein
Director of Engineering, Facebook

ACKNOWLEDGMENTS

- -
- -

This book would not have been possible without the support of a number of wonderful institutions and individuals. First and foremost, we would like to thank all of the students who participated in our workshops, testing out early versions of the tutorials presented here. They were patient, enthusiastic, creative, critical, and vibrant learners and teachers. This book could not have happened without them. Special thanks goes to student Mira Li, who, in one of our interactive monster workshops, designed the monster we use as an example in chapter four.

We would also like to thank the educators and community organizers who helped us recruit participants for our trial workshops. In particular, Susan Klimczak from the Learn to Teach, Teach to Learn Program (L2TT2L) was an exceptional partner—on countless days she helped urban kids with limited resources get to MIT, often in the heart of cold Boston winters, and she always participated in sessions with warmth, generosity, and humor. Ed Baafi, from L2TT2L and ModKit, also spent many weekends working with youth in our lab. Jodi Sandler recruited a terrific group of students from Higgins Middle School and many MIT parents signed their children up for our sessions.

The social scientists and curriculum developers at the National Center for Women and Information Technology (NCWIT) were essential collaborators on this project. Wendy Dubow developed an evaluation framework that we used for our workshops and helped us make sense of the data we gathered. Stephanie Weber and Jane Krauss provided us with tremendously valuable feedback on the tutorials. Without their careful reading and testing, the book would be much harder to understand and use.

Toward the end of our development process, another group of educators generously used our tutorials to teach their own sessions and their comments were invaluable. Debbie Fields and her Craft Technologies class at Utah State University gave us an exciting glimpse into how people could customize and "hack" our tutorials and uncovered critical bugs. Ethan Berman and his teachers at the i2 STEM Summer Camp helped us understand how teachers would use the book in the field.

Tayo Falase was an exceptional editorial and technical assistant on this project. David Mellis, Dia Campbell, Sean Solowiej, Kylie Peppler, Eric Lindsay, Jennifer Jacobs, Hannah Perner-Wilson, Jie Qi, Sam Jacoby, Nancy Buechley, Larry Buechley, Evan Buechley, Yasmin Kafai, Jeffrey Lin, Ali Wyne, and Debbie Fields also gave us excellent editorial feedback and advice.

Funding from the National Science Foundation, through a grant from their Broadening Participation in Computing program (BPC-0940520), made the research that this project is based on possible. (Note: Any opinions, findings, and conclusions or recommendations expressed in this material are those of the authors and do not necessarily reflect the views of the National Science Foundation.) Additional funding from MIT and the MIT Media Lab allowed us to compile and design this volume.

INTRODUCTION

Welcome to *Sew Electric* and the marvelous world of electronic textiles! This book will show you how to make your own soft, colorful, and wearable electronics. You'll play with fabric, light, and sound to build a glowing bracelet, a singing stuffed monster, a fold-up bookmark, and a fuzzy cloth piano. Along the way, you'll learn how to sew, design electronics, and write computer programs.

The cutting edge field of electronic textiles or e-textiles is a recent development in both design and engineering. E-textiles are fabrics with embedded electronics, including sensors, lights, motors, and small computers. Designers of e-textiles keep things soft by employing new materials like conductive thread, conductive fabric, and flexible circuit boards. E-textiles are used in many different domains and settings including astronaut space suits, wearable medical devices, and haute couture fashions. You'll find them in Lady Gaga costumes that change shape, high-tech military tents, and jackets that keep you warm while you snowboard.

This book is designed to provide a creative, hands-on introduction to this fascinating new field. In it, you'll learn how to use the LilyPad Arduino toolkit to design and build your own e-textile projects.

THE PROJECTS

The five main chapters of the book contain step-by-step instructions for making four projects: a cloth bookmark, a sparkling bracelet, an interactive stuffed monster, and a soft fabric piano. Each tutorial is a self-contained lesson. You may want to do just one of the projects, or all four. Each tutorial builds on the previous ones though, so it's a good idea to review the early tutorials before you start on one of the more advanced projects.

The book begins with the bookmark tutorial, which describes how to build a cloth bookmark with a stitched-in LED light that you can use to read after dark. This tutorial introduces basic sewing and circuit design techniques and explores some fundamental ideas in electronics. You can build and decorate a bookmark in about 2-4 hours.

In the sparkling bracelet tutorial, you'll make a glowing bracelet with an LED that flickers, fades, or blinks—your choice. You'll experiment with a small sewable computer chip called the LilyTiny and learn how programmable components like the LilyTiny let you create more interesting and complex behaviors in e-textiles. This project should take you 3-5 hours to complete—a full afternoon.

The programming tutorial explains how to create your own blinking and flickering patterns for LEDs by writing Arduino programs for the LilyPad Arduino, another small computer chip. This chapter introduces the Arduino programming software and provides step-by-step instructions for writing Arduino programs and then loading them onto the LilyPad Arduino. This tutorial sets the foundation for the next two projects and will take you around 3-5 hours to work through.

In the monster tutorial you'll build a cute and cuddly monster (or a fierce and ferocious one depending on your design) who sings and twinkles when you squeeze its paws. This project brings together everything you've learned in the previous chapters—programming, sewing, electronics, and design. Since this is an in-depth craft and engineering project, it will take longer to build. Budget 4-10 days to design, sew, and program your monster.

The final tutorial guides you through the process of making a soft fabric piano—a musical instrument made out of cloth. The piano will play one kind of sound when it's connected to a computer and another when it's unplugged. This project dives into programming more deeply. It focuses especially on how e-textiles can communicate with and control your computer in different ways. This project should also take 4-10 days to build.

TROUBLESHOOTING GUIDES, MATERIALS REFERENCE, AND GLOSSARY

You'll probably make a few mistakes as you build your projects—everyone does! The troubleshooting flow charts in each chapter will help you find and fix these problems. The charts are organized into electrical problems and programming (or "code") problems. They are designed to help you identify the sources of errors. Techniques for fixing these errors are described in detail in the Troubleshooting Solutions section of the book. Once you identify the cause of an issue with a flow chart, follow its page reference to the Troubleshooting Solutions section for detailed information on how to fix the problem.

The Materials Reference section provides detailed descriptions of each material, tool, and component used in the projects. The descriptions include information on where to purchase supplies and how much they're likely to cost. Where it is possible to use an alternative part or material, the Materials Reference also describes these options.

The Glossary contains definitions for the uncommon and technical terms used in the book. As you are reading, you will encounter **bold words**. Definitions for these words can be found in the Glossary. If you encounter a bold word and you're not sure what it means or have never seen it before, look it up there.

WEBSITE

You can find all of the material in this book, along with updates and downloads on the book's website: http://www.sewelectric.org. Visit the site to copy and paste code examples, find the latest information on materials and supplies, and see more project ideas.

HAVE A WONDERFUL TIME DESIGNING AND BUILDING YOUR E-TEXTILE PROJECTS!

BOOKMARK BOOK LIGHT

In this first **e-textile** project, you'll explore how to make soft electronic circuits using conductive thread, LEDs, batteries, and fabric. You'll stitch together a fuzzy bookmark that you can use to read in the dark.

Time required: 2-4 hours

Chapter contents

Collect your tools and materials	p. 6
Design your bookmark	p. 8
Build your bookmark	p. 10
Troubleshooting	*p. 17*
Understanding your circuit	p. 18
Decorate your bookmark	p. 21
Experiment with extensions	p. 22

COLLECT YOUR TOOLS AND MATERIALS

This project uses a basic set of electronic, sewing, and sketching supplies. You can find detailed descriptions of all of these tools and materials in the reference section, which begins on page 168.

ELECTRONIC MATERIALS:

Coin cell battery holder

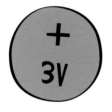

Coin cell battery

LilyPad LED

Conductive Thread

CRAFT MATERIALS & TOOLS:

Chalk or pencil for marking fabric

Felt

Paper

Glue

Large-eyed Needle

Colored pencils for drawing design sketches

Scissors

DESIGN YOUR BOOKMARK

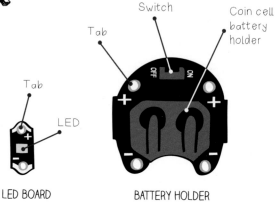

CHECK OUT YOUR ELECTRONIC PIECES

Get out your **LED** (short for "Light-Emitting Diode") board and battery holder and look at them closely. The battery holder has four silver-rimmed holes, called **tabs**, that can be sewn through. Next to each tab is a (+) or (-). The two (+) tabs are connected to the (+) side of the battery that slides into the holder and the two (-) tabs are connected to the (-) side of this battery. The LED board also has two tabs labeled (+) and (-). These are connected to the (+) and (-) sides of the LED.

LED BOARD   BATTERY HOLDER

A NOTE ABOUT WASHING ELECTRONICS

All of the pieces in this project, except the coin cell battery, are washable. If your project gets grungy, take your battery out of its holder and then wash the project in cold water with a mild detergent. Let it drip dry.

CIRCUIT DESIGN OVERVIEW

You're going to stitch together a **circuit** with your pieces that looks like the drawing to the right. The LED will glow when electricity flows from the (+) side of the battery, through the conductive thread, through the LED and back to the (-) side of the battery.

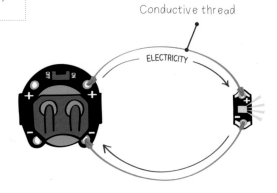

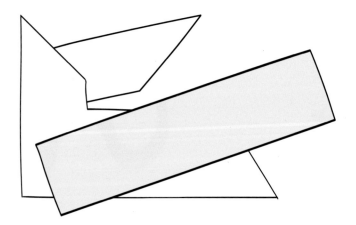

BOOKMARK DESIGN

Decide on a size and shape for your bookmark and draw its outline on a piece of paper. It should be at least 2" (5cm) wide so that your battery holder fits on it easily. Cut out this outline. You'll use it as a template for cutting out your fabric later.

On a second piece of paper, use your template to draw a new bookmark outline. Put your template aside.

Decide on a color scheme and a decorative theme for your bookmark and add these details to your sketch. Figure out where you want your battery holder and LED to be. Draw these on the sketch too.

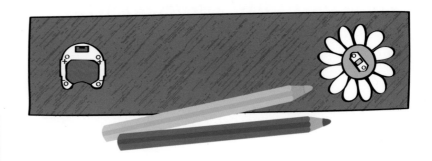

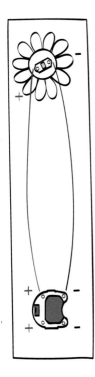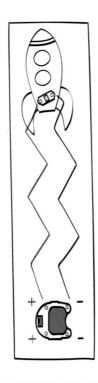

CIRCUIT DESIGN

Plan out the conductive thread connections between the components. In electronics terminology, each of these connections is called a **trace**. In your circuit, there are two traces: the (+) trace that connects the (+) side of the battery holder to the (+) side of the LED and the (-) trace that connects the (-) tabs of the battery holder and LED.

To the left are a few design examples. (The rest of this tutorial will use the first one.) Notice how the (+) traces are drawn in red and the (-) traces are drawn in black. These are the traditional colors used to indicate (+) and (-) in electronics. Use your colored pencils to draw the electrical connections on your design sketch.

Things to keep in mind when you design circuits:

- It's important to connect a (+) tab on the battery holder to the (+) side of the LED, and a (-) tab on the battery to the (-) side of the LED. If you connect your LED backwards it won't light up. LEDs and batteries have what's called **polarity**. This means that the (+) and (-) sides are different and they need to be connected in the correct orientation for a circuit to function.

- If the (+) and (-) traces touch each other this creates what is called a **short circuit**. (You'll learn more about this momentarily.) Short circuits will drain your battery and prevent your LED from turning on. You want to keep the (+) and (-) traces as far away from each other as possible.

- Traces can form part of your decoration. They can be curvy or square, zigzags, loop-d-loops, or straight lines. Alternatively, they can be covered by decorations. Be creative in your layout.

BUILD YOUR BOOKMARK

Use a pencil or a piece of chalk to trace your bookmark template onto your fabric. Cut out this shape.

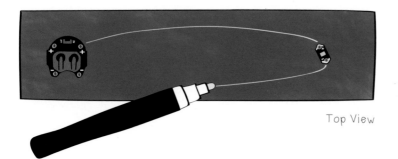

Top View

Glue the battery holder and LED to the top side of the fabric, using your design drawing as a reference. This will keep them in place as you sew. Use only as much glue as you need to attach the pieces and be careful not to fill any of their tabs with glue. You'll need to sew through these in a moment. Note: Do not put the battery in the battery holder yet. You'll do this when you finish building your bookmark.

Using a piece of chalk or a pencil, lightly mark the traces that connect your battery holder and LED on your piece of fabric. This will make it easier for you to create your connections, which you're now ready to start stitching.

Top View

THREAD YOUR NEEDLE

Get out your needle and conductive thread. Cut off 2-3' (1 meter) of conductive thread and thread it through your needle. Pull the conductive thread partway through the needle so that you have one long tail and one short tail coming out of the needle.

TIE YOUR FIRST KNOT

To ensure that your stitching doesn't come undone, you need to tie your thread securely to your fabric before you start sewing. Follow along with the figures on this page and the next as you go through the steps. Note: If you have never sewn before, you may want to practice tying a knot and sewing stitches on a piece of scrap fabric before you start your bookmark.

1

On the underside of your fabric, directly underneath one of the (+) tabs on your battery holder, sew through a small piece of fabric with your needle. Note: Don't sew through the (+) tab itself, just through the fabric.

2

Pull your needle through the fabric so that a short (approximately 1" (2.5cm) tail is left behind.

3

Hold this tail with your other hand and thread the needle underneath the tail to make a small loop.

Bottom View

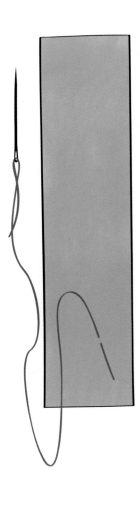

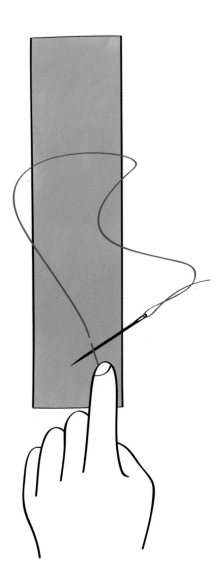

Guide your needle back through the loop you made in step 3. Pull the needle tightly to produce a snug knot on the surface of your fabric.

Guide your needle back through the fabric, creating a new loop.

Guide your needle through this loop.

Pull it tightly to produce the final knot.

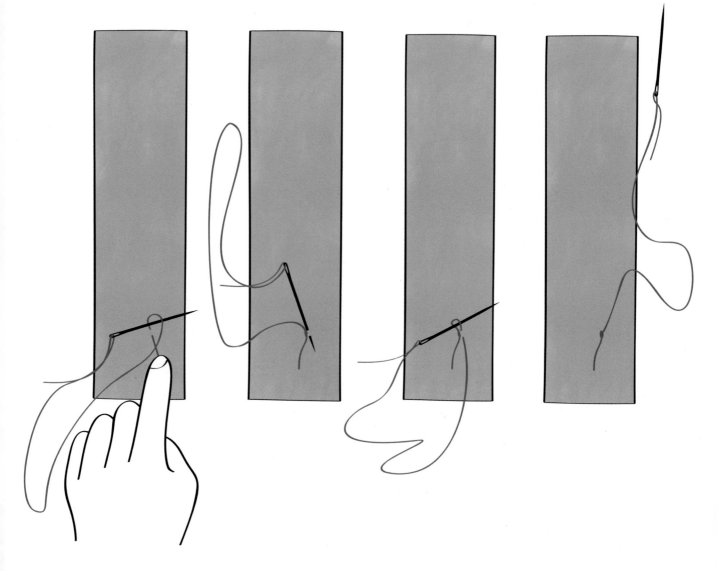

STITCH THE (+) TAB OF THE BATTERY HOLDER

Push your needle up through the fabric right next to the (+) tab of the battery holder. Tug on your needle to make sure you don't leave any excess thread on the underside of your fabric.

Poke the needle through the (+) hole to create a loop around the (+) tab and secure the battery holder to the fabric. Pull the thread through tightly to make a snug connection between the thread and the tab.

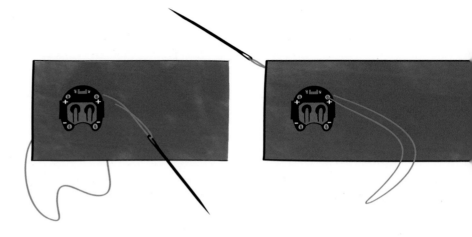

Top View

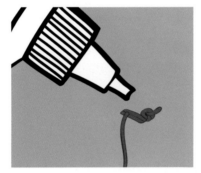

Bottom View

Do this at least two more times to make at least three loops around the tab. Make sure you maintain tight contact between the thread and the tab by tugging on the thread after each stitch.

Before you move on to the next step, put a small dab of glue on the knot on the underside of your fabric to make sure it doesn't unravel. Trim the excess thread on your knot to about ¼" (6mm).

STITCH THE (+) TRACE

Now you're ready to begin sewing the trace from the (+) tab on the battery holder to the (+) tab on the LED. To do this you'll use a simple technique called a **running stitch** that weaves the thread between the front and back sides of the fabric.

Begin with your needle on the underside of the fabric, near the battery holder tab you just sewed. If your needle is on the front side, poke it through to the back side.

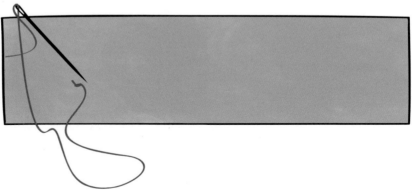

Guide your needle up through the fabric about a ¼" (6mm) away from the battery holder to make your first stitch.

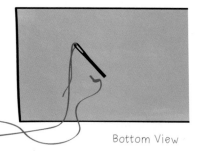

Bottom View

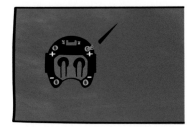

Top View

Then, sew down through the front side of your fabric another ¼" (6mm) toward your LED. Repeat this process until you reach the (+) side of your LED. You should make neat even stitches that follow the line you drew from your battery holder to your LED.

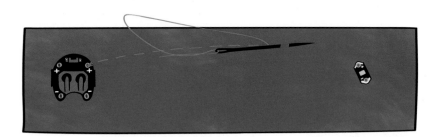

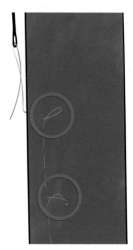
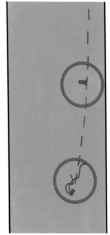

Bottom View

Pause every few stitches to check for loose or tangled threads. Check both the front and back of the fabric to make sure there aren't any tangles or knots hiding on the side you can't see. Also make sure that you're not gathering and puckering your fabric as you stitch. If your fabric is puckered, take time to flatten it back out before you resume stitching.

Once you reach the LED, make at least three tight loops around its (+) tab the same way you attached your battery holder.

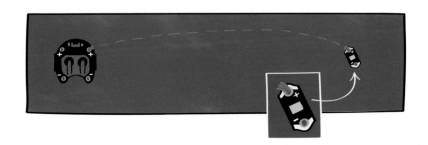

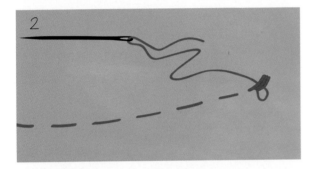

Bottom View

TIE A FINISHING KNOT

Now you need to tie a knot to complete this trace. Begin by making sure your needle is on the back side of the fabric. (You want to hide all of your knots on the back of your bookmark.)

Now you'll repeat the steps you followed to tie your beginning knot:

1. Guide your needle through one of your earlier stitches.

2. This will create a loop of thread.

3. Push your needle back through the loop. Pull on it slowly and firmly until the loop has tightened into a knot.

Repeat this process at least one more time to make a secure knot. Cut your thread, leaving behind about a ¼" (6mm) long tail. Put a dab of glue on the tail to make sure it doesn't unravel.

STITCH THE (-) TRACE

Sewing on the (-) trace is just like sewing on the (+) trace except you'll stitch from a (-) tab on the battery holder to the (-) tab on the LED.

Begin by tying a knot and making at least three tight loops around the (-) tab on the battery holder. (Note: Refer back to page 11 for knot tying instructions.) Trim the tail on your starting knot and put a dab of glue on it to make sure it doesn't come undone.

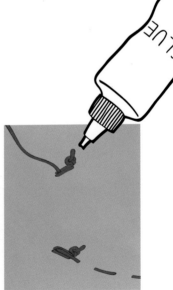

Then, following the line you drew on your fabric, stitch from this tab to the (-) tab on the LED, checking periodically for tangles, loose threads, and puckering. When you reach the (-) tab on the LED, loop through it snugly at least three times before tying a knot, cutting your thread, and securing your knot with a dollop of glue.

As you're sewing, don't let the (-) trace touch or come close to the (+) trace. This is particularly important in places where there are knot-ends or loose threads. When you're finished, double check the front and back of your fabric for loose threads and long knot tails.

In the example on the right, long knot tails are touching and creating short circuits.

Top View

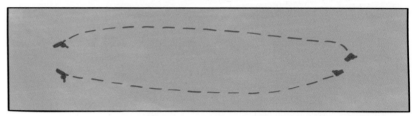

Bottom View

TEST THE CIRCUIT

The circuit is now complete and (hopefully) ready to use! You just have to test it out.

Slide your battery into the holder. The coin cell battery has a flat side with a (+) sign on it that should face up as you put it into the holder.

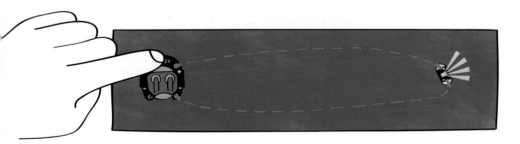

Flip the switch on your battery holder from off to on. Your LED should now shine brightly. If it doesn't, see the troubleshooting section on the next page.

TROUBLESHOOTING

ELECTRICAL PROBLEMS

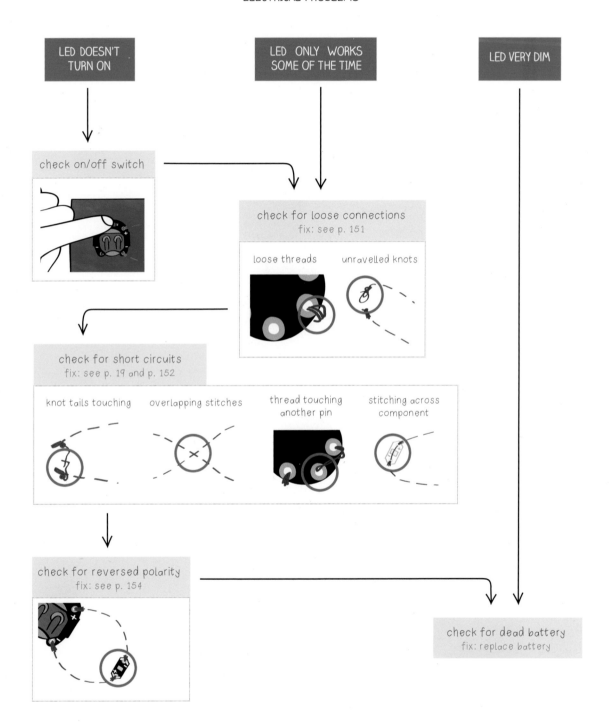

LED DOESN'T TURN ON

LED ONLY WORKS SOME OF THE TIME

LED VERY DIM

check on/off switch

check for loose connections
fix: see p. 151

loose threads   unravelled knots

check for short circuits
fix: see p. 19 and p. 152

knot tails touching   overlapping stitches   thread touching another pin   stitching across component

check for reversed polarity
fix: see p. 154

check for dead battery
fix: replace battery

UNDERSTANDING YOUR CIRCUIT

Now that you've built a circuit, it's time to explore how it works.

Generally, a circuit consists of a power source like a battery that is connected through a **conductive** material—a material that electricity flows through easily—to other electronics like lights, motors, switches, or sensors. In the circuit you built, you used a coin cell battery, an LED light, and a soft conductive thread to make the connections.

CURRENT, VOLTAGE, AND ENERGY

The LED in your circuit turns on when electricity flows through it. This flow is called an **electric current**. Electric current is measured in amps. In your circuit, current flows from the (+) side of the battery through conductive thread, then through the LED, then more conductive thread, and finally back to the (-) side of the battery.

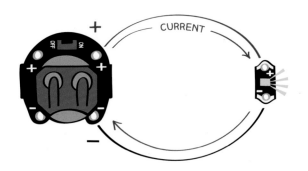

All batteries have two important ratings. They have a voltage rating and an amp-hour rating. The voltage rating tells you how many volts your battery can supply and the amp-hour rating tells you how much current it can supply for how long. Together, these two ratings tell you how much energy is stored in the battery.

The coin cell battery that you're using for this project is a 3-volt battery with an amp-hour rating of .25 amp-hours. This means that the battery can supply .25 amps of current at 3 volts for 1 hour before it dies. The circuit you built uses about .025 amps of current when it's turned on (1/10th of .25 amps). This means that your battery should last about 10 hours. If you used a bigger battery with a higher amp-hour rating—like a camera battery with a .75 amp-hour rating—it would last longer. Here's a table showing how the two compare:

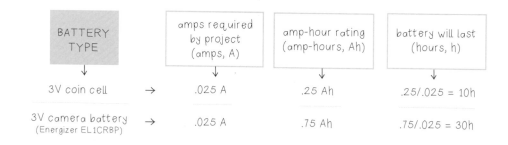

BATTERY TYPE	amps required by project (amps, A)	amp-hour rating (amp-hours, Ah)	battery will last (hours, h)
3V coin cell →	.025 A	.25 Ah	.25/.025 = 10h
3V camera battery (Energizer EL1CRBP) →	.025 A	.75 Ah	.75/.025 = 30h

If you multiply the amp-hour rating by the voltage rating you get a measure of the total amount of **energy** stored in a battery. Energy is measured in watt-hours. The coin cell battery stores .75 watt-hours of energy (3 volts * .25 amp-hours) and the camera battery stores 2.25 watt-hours of energy. In comparison, a typical car battery has a rating of 50-100 amp-hours and can store 1200 watt-hours of energy! See the table on the top of the next page.

Energy ratings for different batteries

BATTERY TYPE	voltage (volts, V)	amp-hour rating (amp-hours, Ah)	energy (watt-hours, Wh)
↓	↓	↓	↓
3V coin cell →	3.0 V	.25 Ah	3 x .25 = 0.75 Wh
3V camera battery →	3.0 V	.75 Ah	3 x .75 = 2.25 Wh
AA battery →	1.5 V	2.00 Ah	1.5 x 2 = 3.00 Wh
car battery →	12.0 V	100.00 Ah	12 x 100 = 1200.00 Wh

The **voltage** rating of a battery tells you the difference in voltage from its (+) and (-) sides. The (-) side of any battery is always at 0 volts. When a battery is fully charged, the (+) side should be at its rated voltage. So, the (+) side of a charged coin cell battery should be at 3 volts, the (+) side of a car battery should be at 12 volts and the (+) side of a AA battery should be at 1.5 volts.

In electronics convention, the (+) side of a battery is called **power** or **HIGH**. The (–) side is called **ground** or **LOW**. Red is the color used to indicate power and black is the color used to indicate ground in electrical diagrams.

Voltage: symbols, values, and names for a 3V battery

symbol	voltage	other name
+	3 volts	power, HIGH
-	0 volts	ground, LOW

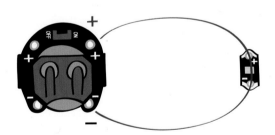

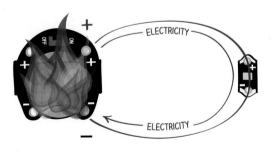

SHORT CIRCUITS

What would happen if you connected the (+) and (–) sides of your battery directly? That is, what if you sewed conductive thread right across the LED's (+) and (–) tabs?

Well, imagine connecting the (+) and (–) sides of a car battery together. Or, even worse, connecting the two slots in an electrical outlet on your wall. When the (+) and (–) sides of a power source like a battery or an electrical outlet are directly connected to each other with a conductive material—like a jumper cable, a bobby pin, or a piece of conductive thread—this is called a **short circuit** or simply a **short**. When a short circuit occurs, the circuit's power supply releases a tremendous burst of energy. When this happens in a battery powered circuit, most of the energy that is stored in the battery is released all at once. If you "short circuit" your coin cell battery, the entire .75 watt-hours will be released in an instant.

A burst like this can be very dangerous and destructive. It is likely to ruin your battery, since it drains all of the battery's energy very quickly, and can electrocute or burn you if the power supply is powerful enough (like a car battery or a wall outlet).

Fortunately, the batteries you're using for the projects in this book won't shock or burn you, even if you create a short circuit. But, if you do create one, your project will not work and you'll quickly ruin your battery.

Short circuits are easy to create in fabric circuits. A loose thread, unravelling knot, messy stitch, or folded piece of fabric can cause your battery's (+) and (−) to touch each other.

It's important to think about and check for short circuits as you design, sew, and troubleshoot your projects. Examples of problems that may cause short circuits are shown below. From left to right: a loose thread brushes against a circuit board, two traces cross, and long knot tails come into contact. See page 152 for more information.

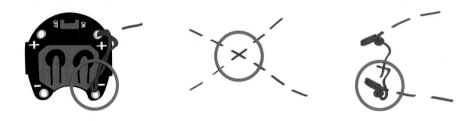

SWITCHES

Electric current only flows through a circuit when there is a complete path leading from (+) on the battery through the circuit and back to (−) on the battery; from power to ground. If there is a break in this path, the electricity stops moving. If you patch the break in the path with a conductive material, the electricity is able to flow again.

This is how switches work. A **switch** opens up a break in a circuit and then closes it. When you flip a light switch on your wall this is what it's doing. This is also what is happening when you flip the switch on your battery holder.

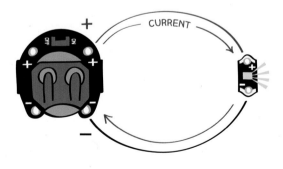

Switch closed

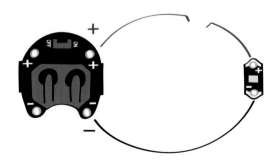

Switch open

DECORATE YOUR BOOKMARK

Now that your circuit is complete and you understand how it's working, it's time to decorate your bookmark. Following your original design, glue and sew on additional decorations to make your bookmark beautiful and personal.

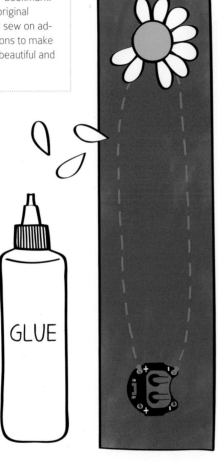

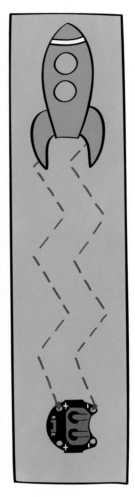

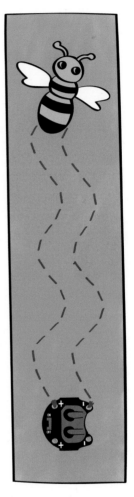

READ

Do some midnight reading with your new glowing page holder!

WASH

You can wash your bookmark in cold water with a gentle detergent, but take the coin cell battery out of its holder first and don't put your bookmark in the dryer.

EXPERIMENT WITH EXTENSIONS

ADD MORE LEDS

If you want to add extra bling to your bookmark, you can sew in a second LED. (If you're so inclined you could even sew in a third, a fourth, a fifth, and so on...) To add additional LEDs to your circuit, you need to stitch them on in the electrical configuration called parallel. A **parallel circuit** with two LEDs (sewn on "in parallel") looks like the diagram on the right.

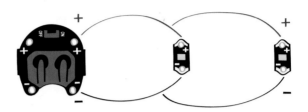

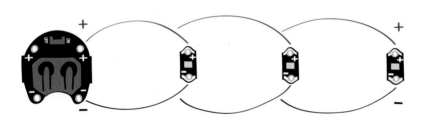

A circuit with three LEDs sewn on in parallel looks like the diagram on the left.

Note that in each of these diagrams all of the (+) tabs on the LEDs are sewn together and all the (-) tabs are sewn together.

To add an LED to your project, stitch the (+) tab of the new LED to the (+) trace of the original circuit and the (-) tab of the new LED to the (-) trace of the original circuit.

Where the new (+) trace intersects the old (+) trace, loop the new trace's thread several times around the old trace's thread before tying a knot, trimming it, and securing it with glue. Do the same at the point where the new (-) trace intersects the old (-) trace.

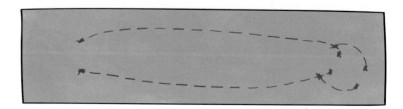

OTHER EXTENSIONS

You could also add a switch to your circuit, or build a new project that includes a switch. You could experiment with building a switch made out of interesting materials like metal beads, metal screws, or paper clips.

Here's one example. The images below show a circuit that includes a tilt switch made from a string of glass and metal beads. The string of beads forms part of the (+) trace of the circuit. A metal bead at the end of this string brushes up against a conductive thread patch that's sewn to the (+) end of an LED to complete the circuit. As this construction tilts and moves, you'll get a lovely flickering effect from the LED. This switch was designed by artist and engineer Hannah Perner-Wilson.

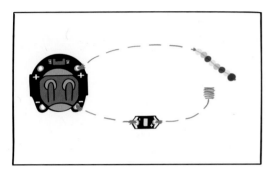

Here are a few more things to think about...Can you turn your bookmark into a bracelet? Or a hair clip? Or a wallet? Can you make your own flashlight, night light, or headlamp?

Now that you know how to sew circuits you can stitch them almost anywhere—on your family's throw pillows, on your backpack, or on your T-shirts. Explore the possibilities!

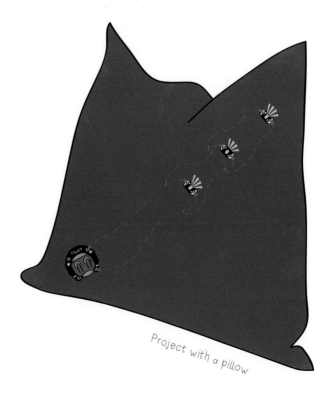

Project with a pillow

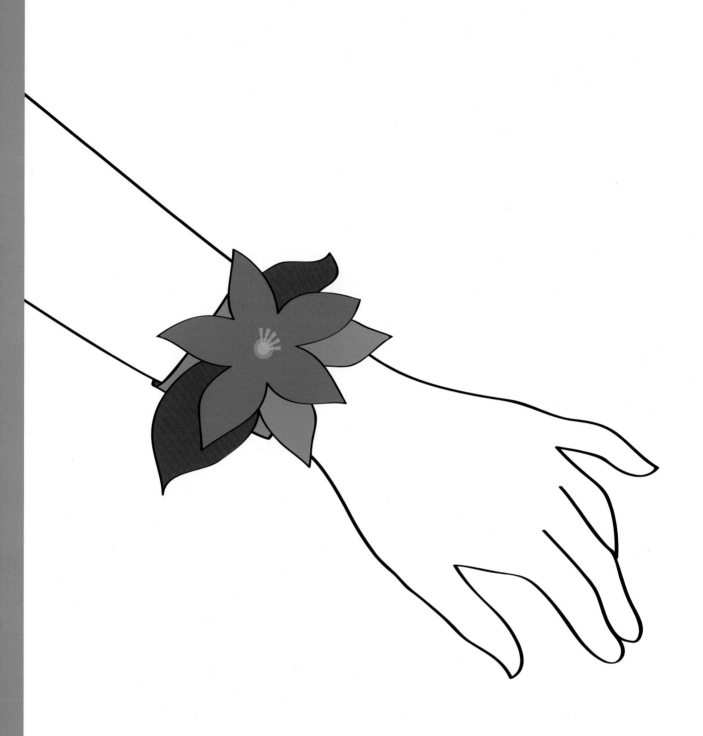

SPARKLE BRACELET

In the last tutorial, you built a simple circuit with a battery and an LED. In this project you're going to begin to explore circuits that include small computers. Tiny sewable computer chips can make LEDs much more fun. Here you'll use one to make an LED blink, fade, or flicker.

Time required: 3-5 hours

Chapter contents

Collect your tools and materials   p. 26
Design your bracelet   p. 28
Build your bracelet   p. 32
Troubleshooting   *p. 35*

A programming preview   p. 36
Decorate your bracelet   p. 38
Experiment with extensions   p. 39

COLLECT YOUR TOOLS AND MATERIALS

This project uses a LilyTiny to control an LED sewn onto a felt bracelet. You'll need a basic set of sewing and sketching supplies in addition to your electronics. Remember that you can find detailed descriptions of all tools and materials in the materials reference section on pages 168-171.

ELECTRONIC MATERIALS:

Coin cell battery

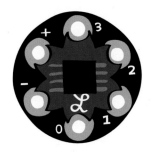

LilyTiny

LilyPad LED

Conductive Thread

Coin cell battery holder

CRAFT MATERIALS & TOOLS:

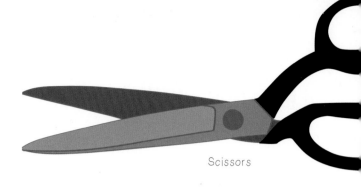

Scissors

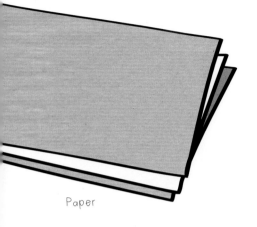

Paper

Large-eyed Needle

Glue

Thread (non-conductive)

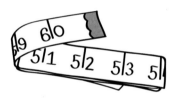

Flexible tape measure
(or a piece of string and a ruler)

Chalk or pencil for marking fabric

Colored pencils for drawing design sketches

2 male snaps

2 female snaps

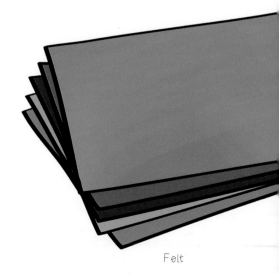

Felt

DESIGN YOUR BRACELET

THE LILYTINY

Before you start designing, it's useful to know a bit more about the LilyTiny—a tiny computer in a sewable package. Take it out and look at it. The small black square in the center of the LilyTiny is a computer chip called a **microcontroller**. It looks a little like a spider with its eight legs. These legs are called **pins** in electronics terminology.

Notice that the LilyTiny board has six silver **tabs** labeled (+), (-), 0, 1, 2, and 3. Each one of these tabs is connected to one of the pins on the chip. If you look closely at the board you can see the conductive **traces** that make these connections. Because tabs are generally attached to pins, we'll refer to pins and tabs interchangeably throughout this volume.

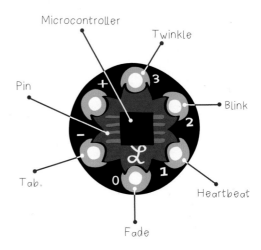

The LilyTiny has been programmed so that an LED attached to each tab will behave differently. These behaviors are labeled in the diagram of the LilyTiny above. If you attach the (+) tab of an LED to the tab labeled 0, the LED will gently fade in and out. If you attach it to tab 1 it will thump on and off in a heartbeat pattern. If you attach it to tab 2 it will blink steadily. Finally, if you attach it to tab 3 it will twinkle like a candle. (Note: If you attach it to the (-) tab it won't turn on at all and if you attach it to the (+) tab it will stay on all the time.) An example behavior and connection for tab 2, the blink tab, is shown in the diagrams below.

You're probably familiar with some of these LED behaviors from your day-to-day life: the fading LED on a sleeping Apple computer, blinking Christmas lights, and LED "candles". In each of these cases there's a tiny computer like the LilyTiny generating that behavior.

The moral of the story here is that the computer chip on the LilyTiny lets you make more interesting and dynamic projects than you could make with just an LED and a battery. This gives you a little glimpse into the power of computers—a topic we'll continue to explore throughout this book. Keep these LED behaviors in mind as you work on the design for your bracelet.

Tab 2: Blink, what you see

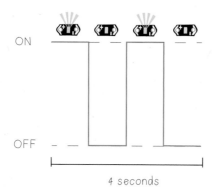

Tab 2: Blink, layout

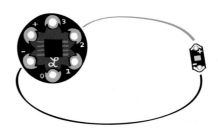

BRACELET DESIGN

The first thing you need to do is figure out how long your bracelet should be. Get out the tape measure, a pencil, and a sheet of paper. Wrap the tape measure around your wrist and write down the measurement. (Note: If you don't have a flexible tape measure, wrap a piece of string around your wrist and mark where one end touches the other. Then use a ruler to measure this length of string.)

Your bracelet will need to be at least 2.5" (6cm) longer than this measurement—this will give you room to attach snaps. The model for the bracelet shown in this tutorial had a wrist that's 7" (18cm) around, so the bracelet shown here will be 9.5" (24cm) long.

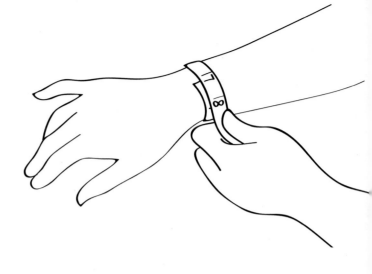

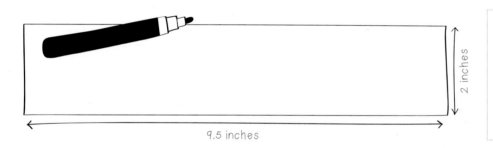

On a piece of paper, draw the outline for your bracelet. It should be at least 2" (5cm) wide so that all of the electronics can fit on it comfortably.

Cut out your bracelet outline to use as a template. On a fresh piece of paper, use the template to create a new bracelet outline for your design drawing. Put the bracelet template aside. You'll use it again in a moment.

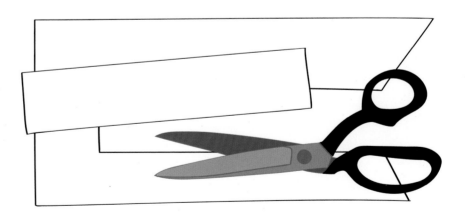

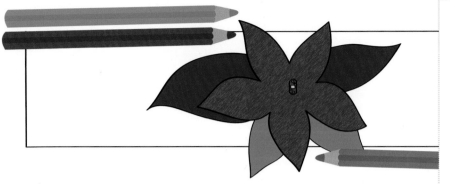

Decide on a color scheme and look for your bracelet. Add these elements to your sketch. Think about which LED behavior you want, where you want your LED to be, and whether you'd like to hide it with decoration or leave it exposed. In the example here, the LED is covered with a pink flower.

Place the battery holder on one side of the bracelet about ½" (10mm) from the edge. Design your closure and mark out where your snaps will go on your design. You should have two sets of snaps: two male snaps and two female snaps.

Male snaps

Female snaps

Male snaps

Female snaps on other side

To make a closure that covers the battery holder, the male snaps should be attached just in front of the battery holder, in between the battery holder and the rest of your design. The female snaps should be attached to the other end of the bracelet on the other side of the felt.

When the bracelet is closed, a flap of fabric will cover the battery holder.

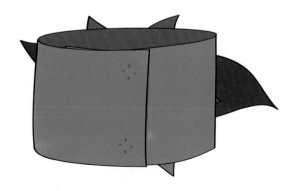

CIRCUIT DESIGN

You've decided where you want to put your LED. Now decide on the placement for your LilyTiny. You want to keep the LilyTiny close to the battery so that it'll be easy to sew. Mark the LilyTiny location on your design sketch.

Next, you want to plan out how you're going to connect the components to each other. You can make your electrical diagram on your bracelet design sketch, or make a new drawing for the electrical elements. Do whatever you think will be most helpful. This page will show the electrical design on its own.

LILYTINY will be
covered with a green leaf.

Draw the connections between the battery holder and the LilyTiny. Connect one of the (+) tabs on the battery holder to the (+) tab on the LilyTiny and one of the (-) tabs on the battery holder to the LilyTiny's (-) tab. As in the bookmark project, (+) connections should be drawn in red and (-) connections should be drawn in black.

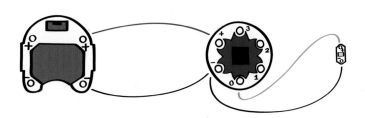

Sketch the connections for your LED. Draw the connection between the (-) side of your LED and the (-) tab on the LilyTiny in black. Then, draw the connection between the (+) side of your LED to the behavior tab you've chosen in a different color. In this tutorial, the LED's (+) tab is attached to tab 0, but you can choose tab 0, 1, 2, or 3. The LilyTiny section at the beginning of this chapter describes the behavior of each tab.

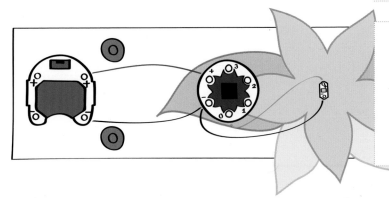

At this point, you may be faced with crossing traces or a complex electrical layout. If this is the case, adjust your design by moving your components and re-drawing your connections. You may also want to change your decorations to hide or show off connecting traces and electronic components. Edit your design until you have something that you're happy with.

BUILD YOUR BRACELET

Using the template you created at the beginning of the design process, mark the shape of your bracelet on a piece of felt with a pencil or a piece of chalk. Cut out this shape.

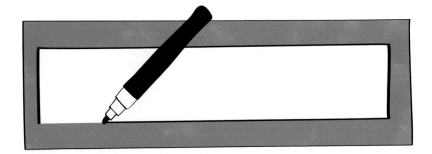

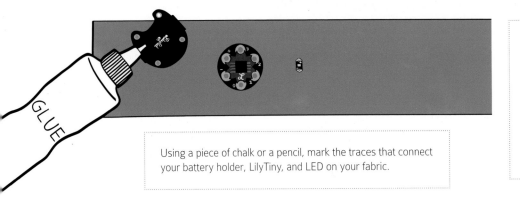

Using a piece of chalk or a pencil, mark the traces that connect your battery holder, LilyTiny, and LED on your fabric.

Glue your battery holder, LilyTiny, and LED to the top side of your bracelet fabric. Use only as much glue as you need to attach the pieces. Be careful not to fill any of the tabs on the pieces since you'll need to sew through these in a moment.

SEW YOUR BATTERY HOLDER AND LILYTINY

Measure out 2-3' (1 meter) of conductive thread and thread your needle. Tie a knot underneath the appropriate (+) tab on the battery holder. (See page 11 in the bookmark tutorial for a detailed description of how to tie a knot.) Push your needle up through the fabric alongside the (+) tab. Sew tightly through the (+) tab at least three times.

Sew from this (+) tab to the (+) tab on the LilyTiny, following the trace you drew. When you reach the LilyTiny, sew through its (+) tab at least three times. When you're done, thread your needle to the back of the fabric—underneath the LilyTiny—and make sure there are no loose threads in your stitching. Pull on the thread until your connections are snug and tie another knot. Trim the tails of both of your knots to about ¼" (6mm) and seal them with dabs of glue.

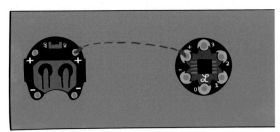

Top View

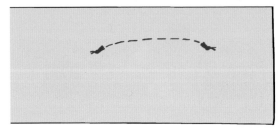

Bottom View

Repeat this process for the (-) tabs on the battery holder and LilyTiny. Sew from a (-) tab on the battery holder to the (-) tab on the LilyTiny following your chalk or pencil lines. Make sure that you sew through each tab at least three times and tie tight knots at each end of your stitching.

Before moving on, trim all of your knots and secure them with glue. Look closely at the stitching you've done so far and make sure that your (+) and (-) threads are not touching each other and creating a short circuit.

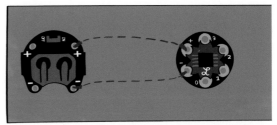

Top View

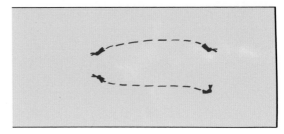

Bottom View

SEW YOUR LED

Tie a knot underneath the (-) tab on the LilyTiny and loop through it snugly at least three times. This thread will connect with the thread you've just sewn for the battery connection. That's fine. (Note: We're going to stop telling you to sew through each tab at least three times now because it seems terribly repetitive, but whenever you sew through a tab, you need to loop through it tightly, three or more times.)

Stitch from the (-) tab on the battery holder to the (-) tab of your LED. Sew through the (-) tab of your LED, securing it tightly, and tie a knot on the underside of your fabric.

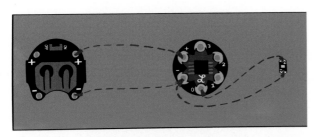

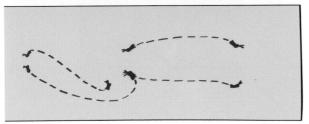

For the (+) tab of your LED, begin by tying a knot underneath the tab you chose on the LilyTiny (tab 0 for the design here). Stitch tightly through this tab and sew across your fabric to the LED.

When you've finished sewing through the (+) tab on the LED, tie a knot underneath it. Tie this knot as far away from the (-) tab's knot as you can to prevent a short circuit. Trim all of your knots to ¼" (6mm) and dab glue on them. Look closely at the stitching you've done so far to make sure that none of your threads are touching where they shouldn't be.

TEST THE CIRCUIT

The electrical part of your bracelet is now done and it's time to test it out. Slide your battery into the holder. The flat side with a (+) sign on it should face up as you put it into the holder. Flip the switch on the battery holder and see what happens. Does your LED blink, flicker, fade, or thump the way it should? If you're having problems, check out the troubleshooting guide on the next page.

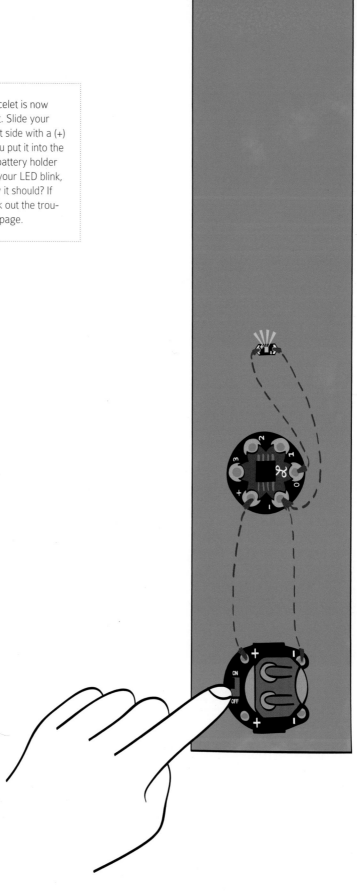

TROUBLESHOOTING

ELECTRICAL PROBLEMS

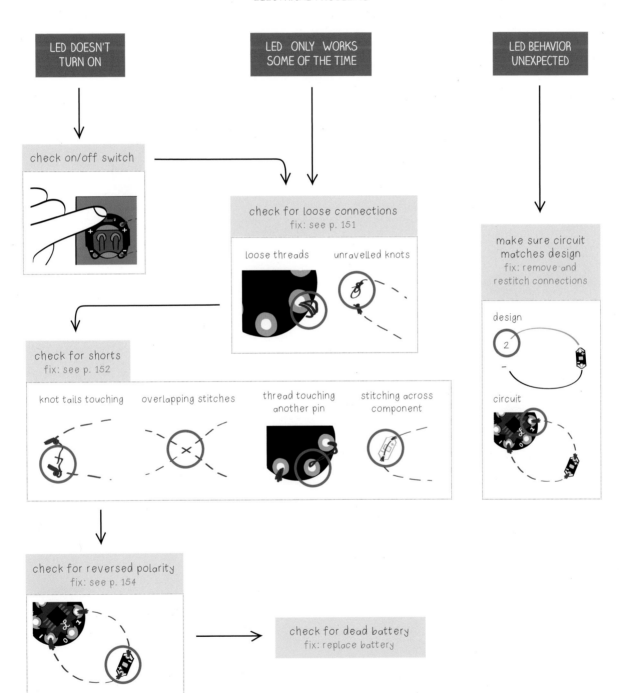

LED DOESN'T TURN ON

check on/off switch

LED ONLY WORKS SOME OF THE TIME

check for loose connections
fix: see p. 151

loose threads   unravelled knots

LED BEHAVIOR UNEXPECTED

make sure circuit matches design
fix: remove and restitch connections

design

2

circuit

check for shorts
fix: see p. 152

knot tails touching   overlapping stitches   thread touching another pin   stitching across component

check for reversed polarity
fix: see p. 154

check for dead battery
fix: replace battery

A PROGRAMMING PREVIEW

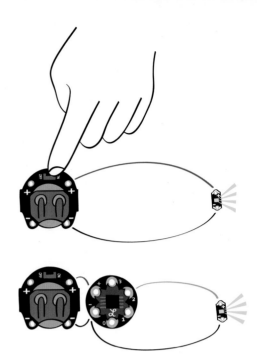

Why is your LED flickering, fading, or blinking? The beginning of this tutorial mentioned that the LilyTiny was programmed to generate this behavior, but what does that mean?

In the circuit from the bookmark tutorial, you could control the LED by flipping the on/off switch on the battery holder with your finger. If you did this very quickly and carefully you could get the LED to blink in a regular pattern, but it'd be hard work.

On the bracelet, you've added a LilyTiny to the circuit. Your circuit is similar to the bookmark circuit except you've sewn your LED to the LilyTiny instead of the battery.

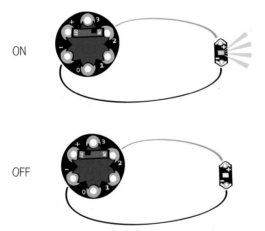

ON

OFF

You can think of each tab on the LilyTiny as a switch that is controlled by the LilyTiny. In the example on the left, an LED attached to pin 2 turns on when this switch is closed and turns off when this switch is open.

The LilyTiny opens and closes these switches in patterns specified by a **program**. What's cool is that these are incredibly fast switches. Each one can open and close millions of times per second in exactly the way the LilyTiny tells it to. What's even better is that the LilyTiny is not just one switch, it's four! Each tab can be programmed with a different pattern.

Think about what these programs might be saying for the different LED behaviors. Start with the simplest behavior, blink—how would you describe the blink behavior if you had to explain it to someone who was flipping a switch? Stop and think about this for a moment and jot down a set of instructions. They should look something like the "program" on the right.

What about the heartbeat pattern? Can you write out a description of that behavior?

blink "program":

```
repeat the following over and over:

    turn the LED ON (close the switch)
    wait for one second
    turn the LED OFF (open the switch)
    wait for one second
```

The fading and flickering behaviors are a bit more complex because the LED appears to gradually get dimmer and brighter. Try writing out a program to describe these patterns.

A rough outline of code for the tab 0 fading behavior is shown on the right and a chart for this behavior, plotting the LED's brightness over time, is shown below.

fade "program":

begin with the LED OFF

repeat the following over and over:

    repeat the following until the LED is fully ON:
      add a little brightness to the LED

    wait for half of a second

    repeat the following until the LED is OFF:
      subtract a little brightness from the LED

    wait for half of a second

Tab 0: FADE, what you see

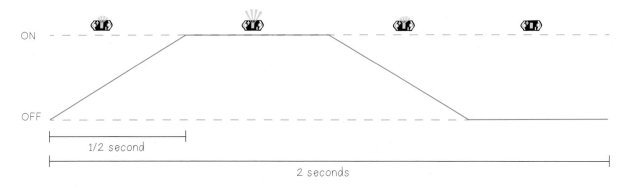

In the next tutorial you'll get to write your own programs to create your own blinking patterns. For now, these examples provide nice insights into why computers are so powerful. Computers can automate annoying repetitive tasks—like flipping a switch over and over. They can work at very high speeds. They're tiny and can be embedded into almost anything—even bracelets. And, most importantly, they're programmable—they do what you tell them to do!

DECORATE YOUR BRACELET

You've finished your circuit and explored how it works. Now it's time to decorate it.

Cut out the decorations you've designed and glue or stitch them to your bracelet. If you're sewing them, use a normal non-conductive thread so that these stitches don't interfere with your circuit.

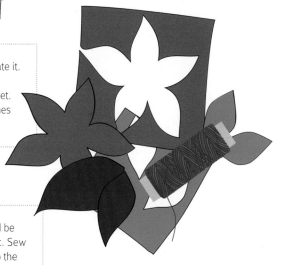

SEW ON YOUR SNAPS

Get out a non-conductive thread to sew on your snaps. Note: Your stitches will be less visible if you choose a thread that is the same color as your bracelet fabric. Sew the two male snaps to the top side of your fabric. Sew the two female snaps to the underside of your fabric. Keep these last stitches especially neat, because they will show up on the outside of your bracelet when you close it.

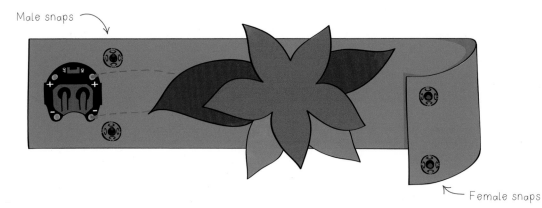

Male snaps

Female snaps

WEAR IT! WASH IT.

Snap your bracelet around your wrist and turn it on. You're ready to dazzle your family and friends, especially when you wear it after dark!

You can wash your bracelet in cold water. Remember to take out the battery first.

EXPERIMENT WITH EXTENSIONS

To make your bracelet even more sparkly, add more LEDs. Add an LED (or more than one) in **parallel** to the one that's on your project.

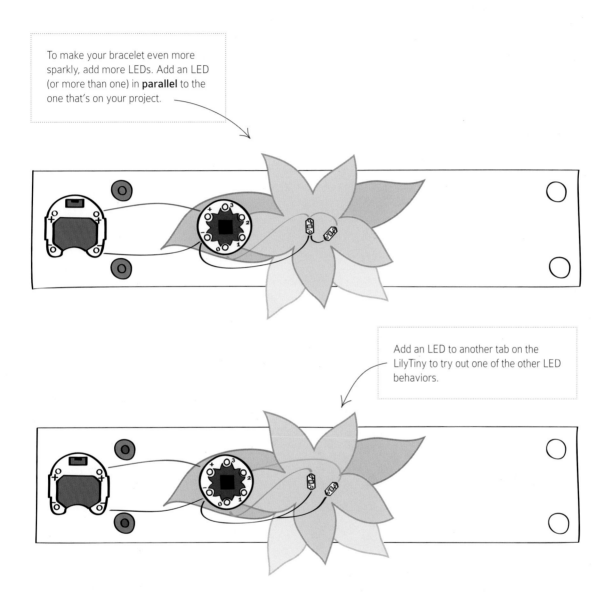

Add an LED to another tab on the LilyTiny to try out one of the other LED behaviors.

Experiment with other ways to extend your project. Can you add a switch to your project? Can you add an LED so that it behaves exactly the opposite of the first LED you added—when the first LED is on this LED is off and vice versa? This is an especially hard challenge. Hint: Think about flipping the **polarity** of your LED.

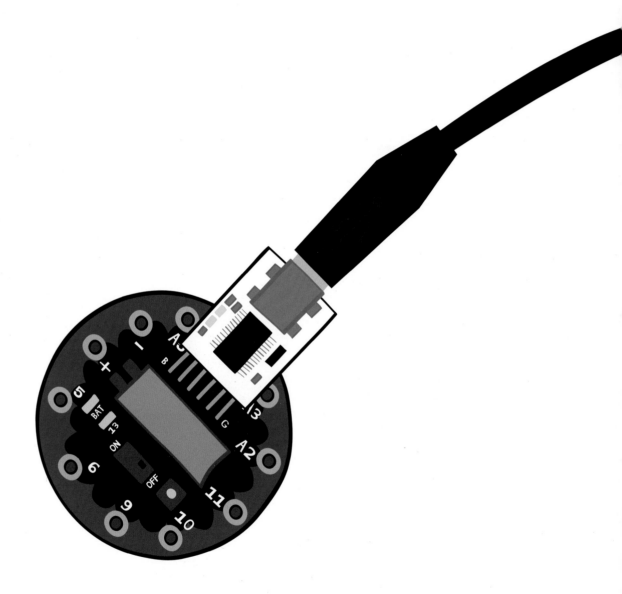

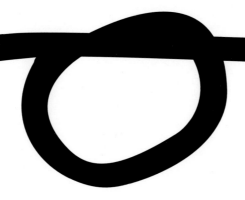

PROGRAMMING YOUR LILYPAD

In this tutorial, you'll learn how to write programs for the LilyPad Arduino SimpleSnap. You'll learn how to make an LED blink and flicker exactly they way you want it to. You'll also begin to explore the ways that programming can make your e-textile projects more interesting, interactive, and personal.

Time required: 3-5 hours

Chapter contents

Collect your tools and materials   p. 42
The Lilypad Arduino SimpleSnap   p. 43
Set everything up   p. 44
Programs and Arduino   p. 46
Basic programming steps   p. 47
Make the LED blink faster   p. 51
Basic code elements   p. 52
Arduino program structure   p. 56
Experiment   p. 58
Troubleshooting   *p. 59*

COLLECT YOUR TOOLS AND MATERIALS

You'll be writing an Arduino program on a computer, transferring that program to a LilyPad Arduino SimpleSnap board (which is also called simply a LilyPad), and running the program on the LilyPad. You'll need a computer, a LilyPad Arduino SimpleSnap, and a mini-USB cable and FTDI board to make connections.

ELECTRONIC MATERIALS:

Mini-USB cable

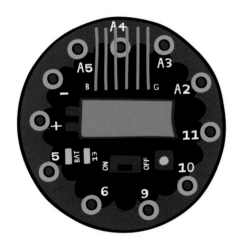

Lilypad Arduino SimpleSnap

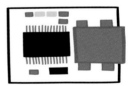

FTDI breakout board

Plus a computer with an
Internet connection

THE LILYPAD ARDUINO SIMPLESNAP

THE BASICS

Before you start programming, it's useful to know a few things about the LilyPad Arduino SimpleSnap—a small computer in a snap-on package. Take out your LilyPad board and take a close look at it. Note: We'll also sometimes call the LilyPad a board, which is short for circuit board. The top side of the board has an on/off switch, a push button switch, a connector that will attach to your FTDI board, a battery, and two LED lights. There are also 11 silver circles on the edge of the board. As in previous tutorials, we'll call these **tabs**. Notice how each tab has a label: +, -, 5, 6, 9, etc.

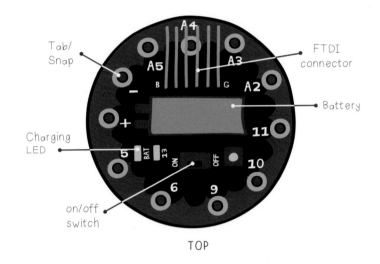

TOP

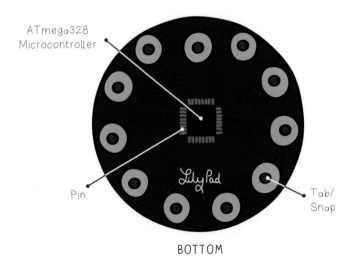

BOTTOM

The bottom side of the LilyPad has a ring of snaps surrounding an assortment of electronics. Each snap is connected to one of the tabs on the top side of the board. The black square in the center of the board is a computer chip called a **microcontroller**. It's similar to the microcontroller on the LilyTiny that you used in the Sparkle Bracelet tutorial. However, this microcontroller, an ATmega328, is larger and more powerful; notice how it has many more **pins** than the LilyTiny's chip. This chip stores programs and controls electronics that are connected to the LilyPad.

THE BATTERY

The LilyPad Arduino SimpleSnap has a built in rechargeable battery. This is the large silver block on the top of the board. When the LilyPad is plugged into your computer the battery will charge. An orange LED on the LilyPad will shine while the battery is charging. When the battery is fully charged this light will turn off.

NOT WASHABLE!

Because it has a built in battery, the Lily-Pad Arduino SimpleSnap is NOT WASH-ABLE! Snap it off of your project before washing your project in cold water with a gentle detergent.

SET EVERYTHING UP

INSTALL THE NECESSARY SOFTWARE

Visit http://arduino.cc/en/Guide/ArduinoLilyPad for the most recent instructions on downloading and installing the Arduino software. Follow the directions for your computer type (ie: PC or Mac) and operating system (ie: Windows 8, Windows NT, Mac OS 10.8, etc.). You'll also need to download and install a piece of softaware called an FTDI driver. Instructions for downloading and installing the FTDI driver are on the same website. Make sure you've installed both pieces of software successfully before moving on. If you have already installed everything and successfully programmed your LilyPad at least once, you can skip ahead to the Programs and Arduino section on page 46.

OPEN THE ARDUINO SOFTWARE

Once you have downloaded and installed the Arduino software, open it by double-clicking the green Arduino icon. If you are using a Mac the software should be in your Applications folder. If you are using a PC it should be in the Arduino folder that you downloaded. Note: If you've already opened the software, close it now, reopen it and follow along with these instructions. When you first open the software you should see an empty window.

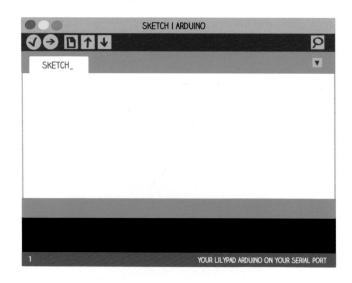

ATTACH YOUR LILYPAD TO YOUR COMPUTER

Attach the FTDI board to your LilyPad. Plug one end of your USB cable into the FTDI board and the other end into your computer. The orange light on your LilyPad may come on. If it does, this means that your battery is charging. If the light stays off, this means that your battery is fully charged.

FTDI breakout board

Mini-USB cable

Lilypad Arduino Simplesnap

SELECT THE APPROPRIATE SERIAL PORT

You need to select a **serial port** so that the Arduino software knows which USB port your LilyPad is attached to. The serial port is the communication channel through which your programs will be sent. To select the correct one, open the Tools → Serial Port menu.

If you are using a Mac, you should see an entry that starts with: "/dev/tty.usbserial- " and is followed by a series of letters and numbers. Select this port.

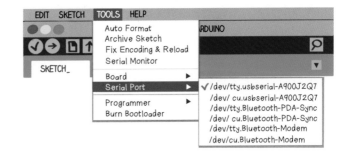

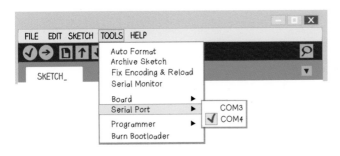

If you are using a PC, the correct serial port is usually the highest numbered "COM" port.

You can also find your serial port by unplugging your LilyPad, looking at the menu, and then plugging your LilyPad back in and looking at the menu again. The serial port entry that has appeared is the one that you want to select.

If you do not see an appropriate serial port entry, this probably means you have not installed the necessary FTDI driver. Visit http://arduino.cc/en/Guide/ArduinoLilyPad for detailed instructions on driver installation.

SELECT THE APPROPRIATE ARDUINO BOARD

Now you need to tell the Arduino software that you are using a LilyPad and not some other Arduino board. Under the Tools → Boards menu, select "LilyPad Arduino w/ ATmega328". Note: This tutorial assumes you are using a LilyPad Arduino SimpleSnap board. If you are using a different LilyPad board, you should select the menu entry for your board.

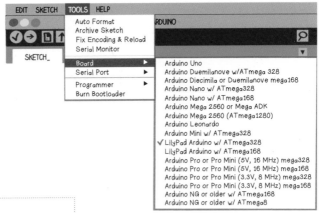

TROUBLESHOOTING THE SOFTWARE SETUP

If you're having trouble installing or setting up the software, see the troubleshooting guide at the end of this chapter or the troubleshooting section of the Arduino website: http://arduino.cc/en/Guide/Troubleshooting.

PROGRAMS AND ARDUINO

Before you get started programming, it's worth taking a moment to reflect on what programs are and what they do. Computer programs are everywhere. Your Internet browser and music player are examples of complex programs. So are Google, Facebook, and the video games that you play at home and online. Programs can also be used to control electronics like lights, motors, and speakers. You can find programs like these in microwaves, cars, robots, and hair dryers.

Computers and programs are powerful and beautiful things that underlie almost every aspect of modern life. Programmed computers carry out boring and repetitive jobs for us, they let us do precise tasks at incredibly high speeds, and they enable us to build dazzlingly complex systems (the Internet for instance). They also enable us to express ourselves in new and dynamic ways.

More formally, a **program** (also called a piece of **code**) is a set of instructions written in a programming language that follows a very precise format. The program does its work when a computer **runs** or **executes** these instructions by following them in order.

THE ARDUINO PROGRAMMING ENVIRONMENT

Open the Arduino software—also called the Arduino programming environment. An empty window will pop up. This window is divided into four sections: a Toolbar, a Code Area, a Status Bar, and a Feedback Area.

Note: Depending on the version of the Arduino software you are using, your window may look different from the one shown below. Don't worry if that is the case. The basic functionality will be the same.

1. Toolbar
The Toolbar provides you with a quick way to do common tasks. Each icon in the Toolbar corresponds to a different action.

 - Compile your code   - Open an existing file

 - Compile and upload your code   - Save the file you're working on.

 - Create a new file   - Open the Serial Port Monitor

Notice that when you hover your mouse over any icon, text pops up that explains what it's for.

2. Code area
The Code area is where you write programs.

3. Status bar
The Status bar gives you information about the status of your program. It tells you when your code is compiling or uploading (these processes are explained in a couple of pages) and lets you know when there's an error that you need to fix.

4. Feedback area
This is where you get feedback about the compiling and uploading processes. If your program cannot compile or the software can't communicate with your LilyPad, you'll get a red error message in this box telling you what went wrong.

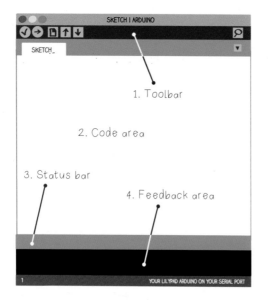

1. Toolbar
2. Code area
3. Status bar
4. Feedback area

BASIC PROGRAMMING STEPS

Now you're ready to start programming! There are four steps to writing a program and running it on your LilyPad:
1. Write the program 2. Compile the program 3. Upload the program to the LilyPad 4. Run the program on the LilyPad.

WRITE THE PROGRAM

You'll begin by trying out pre-written example code that is included in the Arduino software. "Write" is a slightly inaccurate description for the moment, but don't worry, you'll be writing your own programs soon.

Click on the upward pointing arrow in the Toolbar to open Arduino's built-in library of examples. Select 01.Basics → Blink. You can also get a new window with this code by going to the File menu and selecting Examples → 01.Basics → Blink.

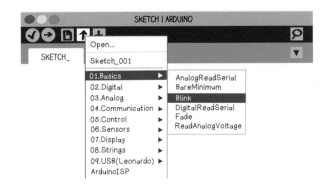

```
/*
  Blink
  Turns on an LED on for one second, then off for one second, repeatedly.

  This example code is in the public domain.
*/

// Pin 13 has an LED connected on most Arduino boards.
// give it a name:
int led = 13;

// the setup routine runs once when you press reset:
void setup() {
  // initialize the digital pin as an output.
  pinMode(led, OUTPUT);
}

// the loop routine runs over and over again forever:
void loop() {
  digitalWrite(led, HIGH);   // turn the LED on (HIGH is the voltage level)
  delay(1000);               // wait for a second
  digitalWrite(led, LOW);    // turn the LED off by making the voltage LOW
  delay(1000);               // wait for a second
}
```

The example program you opened should look like the one on the left.

This program is written in a **programming language** called **C**. The Arduino software only understands programs written in C, so you'll be learning how to write C programs. Assume for a moment that you've written this program. The next step is to compile it.

Note: Depending on the version of the Arduino software you are using, your program may look slightly different. In particular, the areas of grey text (the comments) in your code may be different. Don't worry, your program will work the same way that the example program does. Ignore any differences in your grey text and focus on the other areas of the program.

COMPILE THE PROGRAM

When a program is **compiled** it is translated from the code that you wrote into a new code called **hex code** that the Lily-Pad can understand. The hex code is a very condensed form of the code that you wrote—it is a long string of letters and numbers. An excerpt of hex code is shown on the right.

:1038D00040E350E0225330404040504057FFFACF81
:1038E000962F9F5F692F981728F3909309020895E8
:1038F000982F8091C00085FFFCCF9093C60008955B
:10390000EF92FF920F931F93EE24FF248701809183
:10391000C00087FD17C00894E11CF11C011D111D9A

As you can see, hex code is pretty hard to understand and would be even harder to write! This is why you write C code instead of hex code. If you've made any mistakes in your C code, they will be detected during the compiling phase and you will have to correct them before your code will compile. Only perfectly formatted and punctuated and grammatically correct C code—code with perfect **syntax**—will compile into hex code. To compile your Blink program, click on the check mark button in the Toolbar. When you scroll over this button you should see the word "Verify".

A progress bar that shows how long the compiling process will take will appear in the Status Bar.

If there are no errors in your program, your compilation will be successful and a "Done compiling." message will appear in the Status Bar. The Feedback Area will display a message that tells you the size of your compiled program.

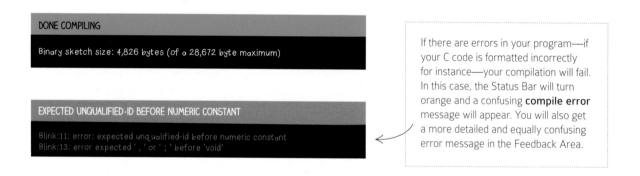

If there are errors in your program—if your C code is formatted incorrectly for instance—your compilation will fail. In this case, the Status Bar will turn orange and a confusing **compile error** message will appear. You will also get a more detailed and equally confusing error message in the Feedback Area.

Most of the error messages that appear in the Status Bar and Feedback Area will be perplexing. They are the computer's way of telling you what's wrong. Don't worry if you don't understand them, but you might be able to get some hints about the error in your program by reading them carefully. As you get more familiar with them they'll become (slightly) more helpful.

Try editing your program to introduce an error. Add a new line of random text to the top of the code. Click the compile button to see what happens. Most likely the error message that appears will be meaningless to you. If you can make sense of it, you are a programming prodigy!

But, Arduino does do one very useful thing when it detects an error. It highlights the line where the error was detected or moves the cursor near where the error was detected.

This can be a good clue about where the problem that you need to correct is located. Keep this in mind as you write and debug your own programs.

Remove the extra line you added and recompile your program before moving on to the next step.

UPLOAD THE PROGRAM TO THE LILYPAD

Once your program has compiled and hex code has been generated, the next step is to **upload** the hex code onto your LilyPad. To upload the compiled code, click on the rightward pointing arrow button in Toolbar. When you scroll over this button you'll see the word "Upload."

After you click it, you'll see messages in the Status Bar telling you that the program is being compiled and then uploaded. If the upload is successful, you'll see a "Done uploading" message in the Status Bar. A message in the Feedback Area will tell you the size of your uploaded program and the amount of memory available on the LilyPad.

DONE UPLOADING

Binary sketch size: 1,108 bytes (of a 30,720 maximum)

1   YOUR LILYPAD

If you have not set up your LilyPad properly, the upload process will fail. The Status Bar will turn orange and an error message will appear. If you receive an error like this, return to the setup section of this tutorial and make sure that you have completed all of the necessary setup steps including installing the FTDI driver and selecting the appropriate serial port in the Arduino software. If you are still having problems, see the troubleshooting chart at the end of this chapter.

PROBLEM UPLOADING TO BOARD.

Binary sketch size: 1,108 bytes (of a 30,720 byte maximum)
avrdude: stk500_recv(): programmer is not responding

RUN THE PROGRAM ON THE LILYPAD

Look at your LilyPad. It should be blinking!

After the Blink program is uploaded, a green LED on the LilyPad should start to flash on and off. It'll turn on, wait a second, turn off, wait a second, and continue doing this until the LilyPad is turned off or reprogrammed.

Blinking green LED

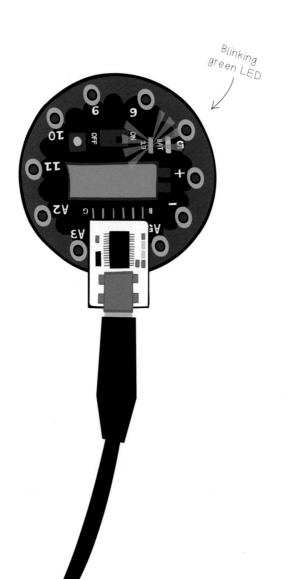

Try unplugging the LilyPad from your FTDI board. Flip the on/off switch on the LilyPad to off. The LilyPad should stop blinking. Now, flip the on/off switch on. The LilyPad should begin flashing again.

When you upload a program to the LilyPad, that program is stored in the LilyPad's **memory**. This means the LilyPad can run the program independently of the computer. Once you've uploaded a program to the LilyPad, you can unplug the LilyPad from the computer and it will remember exactly what you told it to do!

Attach your LilyPad to your computer again so that you can keep programming.

MAKE THE LED BLINK FASTER

Take a closer look at the example code. The four lines of code between the curly brackets highlighted on the right are the heart of the Blink program. When the program runs, these lines execute one at a time in the order that they're written.

- The first line, digitalWrite(led, HIGH);, is the code that turns the LilyPad's LED on.
- The second line, delay(1000);, tells the LilyPad to do nothing for one second (1000 milliseconds).
- The third line, digitalWrite(led, LOW);, is the code that turns the LED off.
- The fourth line, delay(1000);, tells the LilyPad to do nothing for another second.

```
void loop() {
    digitalWrite(led, HIGH);    // turn the LED on
    delay(1000);                // wait for a second
    digitalWrite(led, LOW);     // turn the LED off
    delay(1000);                // wait for a second
}
```

Notice how comments at the end of each line (in grey) describe what the code is doing. Your comments may look slightly different than the ones shown here. Don't worry, your program will still work the same way the example does.

```
void loop() {
    digitalWrite(led, HIGH);    // turn the LED on
    delay(500);                 // wait for half a second
    digitalWrite(led, LOW);     // turn the LED off
    delay(500);                 // wait for half a second
}
```

See if you can change this code to get the LED on the LilyPad to blink faster. To get your LED to blink twice as fast, change each delay(1000); line to delay(500);. Make the changes, compile your code, and upload the new code to your LilyPad.

Experiment with different blinking speeds. What happens when you try a very short delay? Note: the shortest delay time you can use is 1. Decimal numbers will not work here. Upload the code to your LilyPad for each new delay value.

Once you've gotten your LED to blink quickly, see if you can get it to blink very slowly. Next, try getting it to blink in an uneven pattern.

SAVE YOUR CODE

Once you've created a blinking behavior that you like, click on the downward pointing arrow in the Toolbar to save your code. When you scroll over this button you'll see the word "Save". Click "OK" on the popup window that appears. Choose a good name for your file and click on the "Save" button to complete the process.

To make sure that the file saved properly, click on the upward pointing arrow in the Toolbar. This is the "Open" icon in Arduino; when you scroll over it you'll see the word "Open".

The file you just saved should be at the top of the popup menu that appears. All of the programs that you save in Arduino will show up in this menu. This allows you to find and open them easily.

BASIC CODE ELEMENTS

You just did some simple coding. Now you'll explore some of the basic elements of the C language. Reopen the basic Blink example by clicking on the upward pointing arrow in the Toolbar and then selecting 01.Basics → Blink. There are three code elements in the Blink example: **comments**, **variables**, and simple **statements**.

COMMENTS

The first few lines of the Blink example are a **comment**, a piece of text that will be ignored by the computer when the code is compiled.

```
/*
  Blink
  Turns on an LED for one second, then off for one second, repeatedly.

  This example code is in the public domain.
*/
```

You can use comments to make notes—for yourself and others—in your programs. Comments are a useful way to document what a program is for. When you come back to a program long after you wrote it, comments can help you quickly understand what it does. Comments are shown in a greyish brown in the Arduino environment. Notice how the comments in the example code describe what different parts of the program do.

Any block of text between /* and */ characters is a comment. Anything written on a line after two slash characters // is also a comment. The two slash characters can only create comments that are a single line long. You can add comments anywhere and they won't change a program's behavior. Arduino will automatically color them grey. Generally, you should use comments to help you remember what different parts of a program do.

Try adding your own comment to the code and uploading the new program to the LilyPad. The example here shows a single-line comment.

```
// Mary had a little lamb, little lamb, little lamb...
// Pin 13 has an LED connected on most Arduino boards.
// give it a name:
```

Comments can also be useful when you want to temporarily remove a piece of code from your program. For example, if you put // in front of the digitalWrite(led, LOW); line in your program, that line is skipped when you compile and upload the code to the LilyPad. This is called **commenting out** pieces of code. Try making this edit and then compiling and uploading the code. What happens? Look at the new code carefully. Can you see why the LED's behavior has changed?

```
void loop() {
    digitalWrite(led, HIGH);     // turn the LED on
    delay(500);                  // wait for a half second
    // digitalWrite(led, LOW);   // turn the LED off
    delay(500);                  // wait for a half second
}
```

Remove the // you put in front of the digitalWrite(led, LOW); line. Compile and upload your code so that your LED blinks again.

VARIABLES

Writing a program is a lot like writing a cooking recipe. When you write a recipe there is a basic structure that you follow. You list out the ingredients at the beginning of the recipe and you write instructions in the order that they need to be executed. (You don't, for instance, tell a cook to put his pie in the oven before you explain how to make dough for the crust.)

When you write a program you also write instructions in the order that they need to be carried out and list "ingredients"—pieces of code that you'll use in the rest of your program—at the top of your file. A computer, like a good cook, will read and carry out a program in the order it's written.

Variables in a program are like ingredients in a recipe. They are generally listed at the beginning of a program and they identify the components that the program will be using and controlling. After the comment at the top of the program, the next section of the Blink code looks like the block on the right.

```
// Pin 13 has an LED connected on most Arduino boards.
// give it a name:
int led = 13;     // set the variable "led" to the value 13
```

The first two lines in this section are comments. The third line, int led = 13;, creates a variable called led. This line is called a **variable declaration** statement. It declares a variable called led and **initializes** it to 13, identifying the critical ingredient of the Blink program, the LED. The green LED on the LilyPad board is connected to pin 13 on the LilyPad's microcontroller. The line int led = 13; allows you to use the word led instead of the number 13 to refer to the LilyPad's built-in LED in your program. This makes the program much easier to write and understand. You don't need to remember a number, you can use a meaningful name instead.

When you create a variable you set aside a chunk of memory in the LilyPad and give it a name. The int led = 13; line creates a variable called led and stores the value 13 in that memory. There are different **types** of variables in the same way that there are different types of computer files. Computer file types are distinguished by suffixes (ie: .docx, .pdf, .jpg, etc.). Variable types are specified by prefixes when they're created. In the example above, the text int describes what type of variable led is. int stands for **integer**, which means a whole number. Almost all of the variables you will use in your Arduino programs will be of type int.

The form of a variable declaration statement is shown on the right. Any line of code that creates and assigns a value to a variable has this same basic structure.

Variable type	Variable name	assigned to	Value stored in variable
↓	↓	↓	↓
int	led	=	13;

When the program is running, after it's compiled and uploaded, whenever the LilyPad finds the variable led it replaces it with the number 13. What looks like this to you now:

```
digitalWrite(led, HIGH);   // turn the LED on
delay(1000);               // wait for a second
digitalWrite(led, LOW);    // turn the LED off
delay(1000);               // wait for a second
```

Will look like this to the LilyPad when the program is running:

```
digitalWrite(13, HIGH);    // turn the LED on
delay(1000);               // wait for a second
digitalWrite(13, LOW);     // turn the LED off
delay(1000);               // wait for a second
```

To begin to understand why variables are useful, add one to your program. Add a variable called delayTime to your program and set its value to 1000. Add this line right after the int led = 13; line.

Now, replace every delay(1000); line in your program with delay(delayTime);. Also change the comments to correspond to your new code. Compile this code and upload it to your LilyPad. The LED should blink every second, just like before.

```
// Pin 13 has an LED connected on most Arduino boards.
// give it a name:
int led = 13;          // set the variable "led" to the value 13
int delayTime = 1000;
```

```
void loop() {
    digitalWrite(led, HIGH);   // turn the LED on
    delay(delayTime);          // wait for delayTime
    digitalWrite(led, LOW);    // turn the LED off
    delay(delayTime);          // wait for delayTime
}
```

Now try changing the initial value of delay time from 1000 to 500, by editing the int delayTime = 1000; line. Compile this code and upload it to your LilyPad. What happens to the behavior of your LED? Experiment with other values for delayTime, compiling and uploading the code each time you change it.

Notice how you only have to change one line in your program to change the speed of the LED's blinking. Before, when you changed the timing in your code, you had to change at least two lines.

Variables help you keep track of the critical features (ingredients) of your code. When you use a variable, all of the important information about it—its name, type, and value—is listed once at the beginning of the program where it can be easily and quickly changed.

If you would like to save the changes you made to your code, click on the downward pointing arrow in the Toolbar. Click "OK" on the popup window that appears and choose an appropriate name for your file. Click on the "Save" button to complete the process.

SIMPLE STATEMENTS

Begin this next section with a fresh version of the Blink program. Reopen the Blink example by clicking on the upward pointing arrow in the Toolbar and then selecting 01.Basics → Blink.

Statements are essentially computer sentences—lines of code that tell the computer to do something. The line of code digitalWrite(led, HIGH); is an example of a simple statement. All simple statements end with a semicolon (;) in the same way that all English sentences end with a period. Your code is full of simple statements. Here are a few:

```
int led = 13;              digitalWrite(led, HIGH);

                    delay(1000);
```

Each of these lines tells the computer to do something. Each ends with a semicolon. Note that a comment to the right of each statement in your program tells you what each one does.

Try deleting the semicolon at the end of the int led = 13; line. Compile your code. What happens? Do you see an error message?

```
EXPECTED UNQUALIFIED-ID BEFORE NUMERIC CONSTANT

Blink:11: error: expected unqualified-id before numeric constant
Blink:13: error expected ',' or ';' before 'void'
```

When you're writing your own code it's easy to forget to put a semicolon at the end of a statement, but the Arduino software will not compile or upload your code until it is perfectly punctuated.

Errors like missing semicolons can be tricky to find and fix, but Arduino does try to help you. Notice how the cursor in Arduino jumped to the line immediately following the int led = 13; line when you compiled the code. This is a clue about the location of the error. The other clue is in the cryptic message that appears in the Feedback Area: Blink:13: error: expected ',' or ';' before 'void'.

Notice how the message says expected ',' or ';'. It's telling you that the error might be a missing semicolon. You'll slowly learn to make more sense of the strange error messages!

Replace the semicolon at the end of the int led = 13; line and recompile your code.

Next try introducing a different kind of error into your program. In the loop section, delete the W from the first digitalWrite(led, HIGH); statement so that it reads digitalrite(led, HIGH);. Try compiling your code. What happens?

Code needs to be perfectly spelled, capitalized, and punctuated before it will compile! In the case of misspellings and mis-capitalizations, Arduino gives you some extra help. Notice how the text changed from orange to black when you changed the digitalWrite(led, HIGH); statement. Try changing the word LOW on the third line to low. The text color should change from blue to black.

Arduino gives special colors to all of its built-in procedures and variables. When you misspell or mis-capitalize one of these key words, this coloring will disappear. Unfortunately, if you misspell or mis-capitalize any other part of a program—a variable that you created, say—you won't get any color clues to help you find your mistake.

'DIGITALRITE' WAS NOT DECLARED IN THIS SCOPE

```
Blink.cpp: In function 'void loop():
Blink:21: error: 'digitalrite' was not declared in this scope
```

```
digitalrite(led, HIGH);        // turn the LED on
delay(1000);                   // wait for a second
digitalWrite(led, low);        // turn the LED off
delay(1000);                   // wait for a second
```

Experiment with other misspellings and mis-capitalizations, recompiling your code after each edit, to see what kind of feedback you get in different situations. When you're done experimenting, correct your code and recompile it.

```
void loop() {
   digitalWrite(led, HIGH);    // turn the LED on
   delay(1000);                // wait for a second
   digitalWrite(led, LOW);     // turn the LED off
   delay(1000);                // wait for a second
   digitalWrite(led, HIGH);    // turn the LED on
   delay(500);                 // wait for half a second
   digitalWrite(led, LOW);     // turn the LED off
   delay(500);                 // wait for half a second
}
```

To experiment with writing your own statements you're now going to create an uneven blink pattern. Modify the program so that the LED:
- turns on for 1 second
- turns off for 1 second
- turns on for half a second
- turns off for half a second

Add four statements to your code to achieve this behavior.

Compile and upload this new program to your LilyPad and see what happens. Notice how the additional statements change the blinking pattern. Try adding, removing, or editing statements of your own to get different patterns. You might also try experimenting with using variables like delayTime to control these

patterns. Once you've created a blinking pattern you're happy with, click on the downward pointing arrow in the Toolbar to save your changes. Click "OK" on the popup window that appears and choose a name for your file. Click on the "Save" button to complete the process.

ARDUINO PROGRAM STRUCTURE

The previous section looked at different code elements in the Arduino/C language. Now you'll see how they come together into a complete program. Open a fresh version of the Blink example by clicking on the upward pointing arrow in the Toolbar and then selecting 01.Basics → Blink.

Each Arduino program has three main parts:

1. variable declaration section
2. setup section
3. loop section

These sections are analogous to the ingredient list (variable declaration), preparation steps (setup), and cooking steps (loop) in a recipe. The diagram on the right shows where the three parts are located in the code.

When your program runs, it will first define your variables (the ingredients that you need), then execute the setup section once (set everything up to begin cooking), and then execute the loop section over and over (actually do the cooking).

Note: These sections are often preceded by comments that describe what different sections of the program are doing.

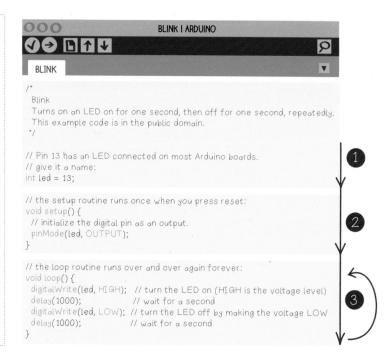

```
/*
  Blink
  Turns on an LED on for one second, then off for one second, repeatedly.
  This example code is in the public domain.
*/

// Pin 13 has an LED connected on most Arduino boards.
// give it a name:
int led = 13;

// the setup routine runs once when you press reset:
void setup() {
  // initialize the digital pin as an output.
  pinMode(led, OUTPUT);
}

// the loop routine runs over and over again forever:
void loop() {
  digitalWrite(led, HIGH);   // turn the LED on (HIGH is the voltage level)
  delay(1000);               // wait for a second
  digitalWrite(led, LOW);    // turn the LED off by making the voltage LOW
  delay(1000);               // wait for a second
}
```

1. VARIABLE DECLARATION SECTION

Programs often begin with introductory comments that explain what the program is doing. These comments come before the variable declarations. It is a good idea to begin every program with comments like these so that when you return to your program later you'll know what it does.

Variables are usually declared at the beginning of a program immediately following the introductory comments. All of the variables that you are using in your code should be listed here, before the setup and loop sections.

2. SETUP SECTION

The setup section, which runs once when the program begins, follows the variable declaration section. Statements that lay the foundation for actions that happen later on in the program are put in the setup section. In particular, pinMode statements are almost always in this section. The section begins with the line void setup() {. Note that all statements in the setup section are placed between an open curly bracket '{' right after void setup() and a closed curly bracket '}' at the end of the section.

```
void setup(){
  // initialize the digital pin as an output.
  pinMode(led, OUTPUT);
}
```

These brackets tell the Arduino software when the setup section begins and ends. Without them, the software is lost and unable to compile your code. Think of these curly brackets as additional punctuation that your code requires. Your code must be perfectly punctuated to compile.

Try removing the closing bracket of the setup section and compiling your code. The error message you get is particularly unhelpful.

```
void setup() {
    // initialize the digital pin as an output.
    pinMode(led, OUTPUT);
```

A FUNCTION-DEFINITION IS NOT ALLOWED HERE BEFORE '{' TOKEN

```
Blink.ino: In function 'void setup()':
Blink:19: a function-definition is not allowed here before '{' token
Blink:24: error: expected '}' at end of input
```

However, it does provide a clue to the problem in the last line: Blink:24: error: expected '}' at end of input This identifies the problem as a missing '}' somewhere in your program.

Replace the closing curly bracket and recompile your code.

Now, move the pinMode(led, OUTPUT); statement to the line after the closing curly bracket. Compile your code. You should get another baffling error message.

```
void setup() {
    // initialize the digital pin as an output.
}
pinMode(led, OUTPUT);
```

EXPECTED CONSTRUCTOR, DESTRUCTOR, OR TYPE CONVERSION BEFORE '(' TOKEN

```
Blink:18: error: expected constructor, destructor, or type conversation before '(' token
```

But, Arduino also highlights the misplaced pinMode(led, OUTPUT); line in yellow—a good clue about the source of your problem.

Return the pinMode(led, OUTPUT); line to its proper position in the setup section and recompile your code.

It's useful to familiarize yourself with these errors because missing or misplaced curly brackets are another common source of problems. Remember that the setup and loop sections should always begin with an open curly bracket and end with a closing curly bracket. (Later on you'll learn that other important blocks of code also begin and end with curly brackets.)

Statements should only appear above the setup section (variable declarations appear here), inside the setup section's curly brackets, or inside the loop section's curly brackets.

If you encounter a compile error, the first things to check for are **syntax errors** like misspellings, missing semicolons, and misplaced or missing curly brackets. See page 155 for more detailed information.

3. LOOP SECTION

After the setup section runs, the loop section runs over and over until the LilyPad is turned off or reprogrammed—hence the word loop. The statements that carry out the main action of your program are in this section.

As with the setup section, statements in the loop section are placed between open and closed curly brackets. These curly brackets tell the computer when the loop section begins (with the opening curly bracket) and when it ends (with the closing curly bracket).

```
void loop() {
   digitalWrite(led, HIGH);    // turn the LED on
   delay(1000);                // wait for a second
   digitalWrite(led, LOW);     // turn the LED off
   delay(1000);                // wait for a second
}
```

```
void loop()
   digitalWrite(led, HIGH);    // turn the LED on
{
   delay(1000);                // wait for a second
   digitalWrite(led, LOW);     // turn the LED off
   delay(1000);                // wait for a second
}
```

You will receive compile errors if either of the curly brackets is missing or if any statements are placed after the closing curly bracket or before the opening curly bracket.

Here are three loop sections that will not compile. See if you can find the problem in each example.

```
void loop() {
   digitalWrite(led, HIGH);    // turn the LED on
   delay(1000);                // wait for a second
   digitalWrite(led, LOW);     // turn the LED off
}
   delay(1000);                // wait for a second
```

```
void loop()
   digitalWrite(led, HIGH);    // turn the LED on
   delay(1000);                // wait for a second
   digitalWrite(led, LOW);     // turn the LED off
   delay(1000);                // wait for a second
```

EXPERIMENT

Now that you have a foundational understanding of Arduino programs and LilyPads, experiment with your LED's behavior. Can you get your LED to Blink in a heartbeat pattern? A seemingly random pattern? Can you get it to flicker like a candle?

Think about some trickier challenges. How could you get the LED to gradually fade on and off? How could you make it blink more and more slowly over time?

Hint 1: Both of these challenges will require additional variables to keep track of blinking speed. Hint 2: You will need to change the values stored in these variables each time the loop section runs.

Remember to save any programs you like by clicking on the downward pointing arrow in the Arduino software so that you can return to them later. You can give each program a different name.

TROUBLESHOOTING

CODE PROBLEMS: COMPILE ERRORS

CODE DOESN'T COMPILE

Compile errors are revealed when you attempt to compile your code and the Arduino software (the compiler) finds a problem. The software then displays an error message in the status bar that may give you a hint about the cause of the error. You will not be able to upload your program until you fix the problem. Common causes of compile errors include missing semicolons, missing curly brackets, misspellings, and mis-capitalizations. See page 155 for more information on compile errors and how to find and fix them.

---

check for missing semicolons ;
fix: add the required semicolon, see p. 156

error examples:

```
int led = 13
```

```
digitalWrite (led, HIGH)
```

your error message may look like:

```
EXPECTED UNQUALIFIED-ID BEFORE NUMERIC CONSTANT
```

```
EXPECTED ';' BEFORE 'DELAY'
```

---

check for missing curly brackets, { and }
fix: add the required curly bracket, see p. 156

error example:

```
void setup() {
  // initialize the digital pin as an output.
  pinMode(led, OUTPUT);
```

your error message may look like:

```
A FUNCTION-DEFINITION IS NOT ALLOWED HERE BEFORE '{' TOKEN
```

```
EXPECTED '}' AT END OF INPUT
```

---

check for misspellings or mis-capitalizations
fix: correct the text, see p. 158

error examples:

```
digitalWrite (led, High);
```

```
dellay (1000);
```

your error message may look like:

```
'DELLAY' WAS NOT DECLARED IN THIS SCOPE
```

---

check for missing parentheses, (and)
fix: add the required parentheses, see p. 157

error example:

```
void loop( ) {
  digitalWrite(led, HIGH);
  delay(1000);
  digitalWrite(led, LOW);
  delay(1000);
}
```

your error message may look like:

```
EXPECTED CONSTRUCTOR, DESTRUCTOR, OR TYPE CONVERSION BEFORE '{' TOKEN
```

---

check for other errors
see p. 155 for more information on compile errors

TROUBLESHOOTING

SOFTWARE PROBLEMS: UPLOAD ERRORS

CODE DOESN'T UPLOAD

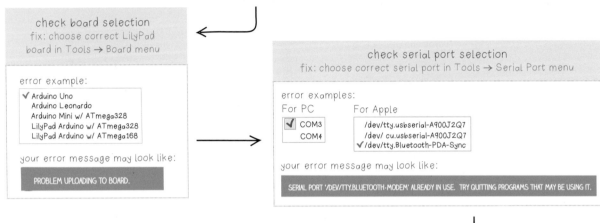

check board selection
fix: choose correct LilyPad
board in Tools → Board menu

error example:
- ✓ Arduino Uno
- Arduino Leonardo
- Arduino Mini w/ ATmega328
- LilyPad Arduino w/ ATmega328
- LilyPad Arduino w/ ATmega168

your error message may look like:

PROBLEM UPLOADING TO BOARD.

check serial port selection
fix: choose correct serial port in Tools → Serial Port menu

error examples:

For PC	For Apple
✓ COM3	/dev/tty.usbserial-A900J2Q7
COM4	/dev/ cu.usbserial-A900J2Q7
	✓ /dev/tty.Bluetooth-PDA-Sync

your error message may look like:

SERIAL PORT '/DEV/TTY.BLUETOOTH-MODEM' ALREADY IN USE. TRY QUITTING PROGRAMS THAT MAY BE USING IT.

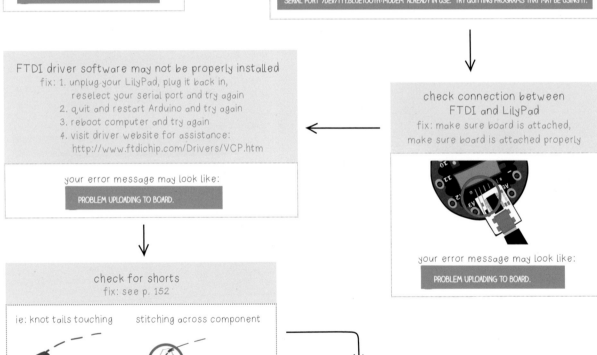

FTDI driver software may not be properly installed
fix: 1. unplug your LilyPad, plug it back in,
     reselect your serial port and try again
    2. quit and restart Arduino and try again
    3. reboot computer and try again
    4. visit driver website for assistance:
     http://www.ftdichip.com/Drivers/VCP.htm

your error message may look like:

PROBLEM UPLOADING TO BOARD.

check connection between FTDI and LilyPad
fix: make sure board is attached,
make sure board is attached properly

your error message may look like:

PROBLEM UPLOADING TO BOARD.

check for shorts
fix: see p. 152

ie: knot tails touching   stitching across component

your error message may look like:

PROBLEM UPLOADING TO BOARD.

check for other issues and get personal help
see the troubleshooting guide on the Arduino website:
http://arduino.cc/en/Guide/Troubleshooting

TROUBLESHOOTING

CODE PROBLEMS: LOGICAL ERRORS

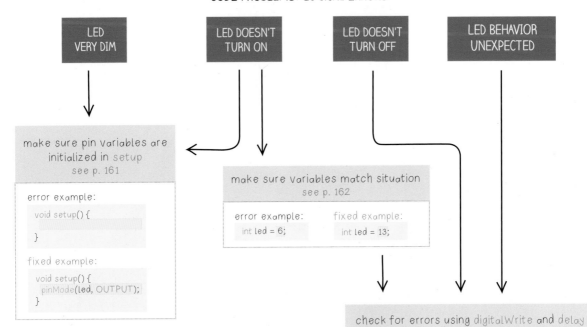

| LED VERY DIM | LED DOESN'T TURN ON | LED DOESN'T TURN OFF | LED BEHAVIOR UNEXPECTED |

make sure pin variables are initialized in setup
see p. 161

error example:
```
void setup() {

}
```

fixed example:
```
void setup() {
  pinMode(led, OUTPUT);
}
```

make sure variables match situation
see p. 162

error example:
```
int led = 6;
```

fixed example:
```
int led = 13;
```

Logical errors occur when your code compiles and uploads, but doesn't behave the way you want it to. These errors are the trickiest to find and fix because the computer doesn't give you any feedback about what might be causing problems, like it does with compile and upload errors.

To fix logical errors, read through your program line by line and try to relate what each line is doing to the behavior you're seeing in your project. The LilyPad does exactly what your program tells it to do; it's misbehaving because you told it to. Your job now is to find the mismatch between what you want it to do and what the program is telling it to do.

In your read through, look especially for missing statements, delays that are too short or too long, and variables with incorrect values. See page 160 for more information about logical errors and a longer list of common problems.

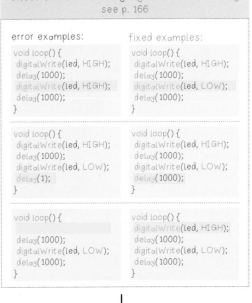

check for errors using digitalWrite and delay
see p. 166

error examples:
```
void loop() {
  digitalWrite(led, HIGH);
  delay(1000);
  digitalWrite(led, HIGH);
  delay(1000);
}
```

fixed examples:
```
void loop() {
  digitalWrite(led, HIGH);
  delay(1000);
  digitalWrite(led, LOW);
  delay(1000);
}
```

```
void loop() {
  digitalWrite(led, HIGH);
  delay(1000);
  digitalWrite(led, LOW);
  delay(1);
}
```

```
void loop() {
  digitalWrite(led, HIGH);
  delay(1000);
  digitalWrite(led, LOW);
  delay(1000);
}
```

```
void loop() {

  delay(1000);
  digitalWrite(led, LOW);
  delay(1000);
}
```

```
void loop() {
  digitalWrite(led, HIGH);
  delay(1000);
  digitalWrite(led, LOW);
  delay(1000);
}
```

check for other errors
see p. 160 for more information on logical errors

LAAAAAA
LALA
LAAAAA

INTERACTIVE STUFFED MONSTER

Now that you know how to sew, put circuits together, and program a LilyPad, you can combine those skills to create a soft interactive project. This tutorial will show you how to design and build a singing and glowing stuffed monster that responds to touch.

Time required: 4-10 days

Chapter contents

Collect your tools and materials	p. 64
Design your monster	p. 66
Begin building	p. 70
Make your monster blink	p. 73
Troubleshooting	*p. 78*
Attach the speaker	p. 81
Make your monster sing	p. 82
Experiment	p. 88
Recreate a complete program	p. 88
Troubleshooting	*p. 90*
Give your monster a sense of touch	p. 92
Troubleshooting	*p. 102*
Sew and stuff your monster	p. 104

COLLECT YOUR TOOLS AND MATERIALS

This project uses a LilyPad Arduino SimpleSnap and an aluminum foil sensor to control an LED and a speaker. You'll need fabric and a basic set of sewing and sketching supplies in addition to your electronics.

ELECTRONIC MATERIALS:

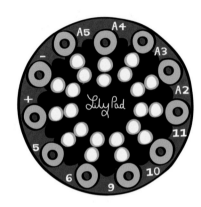

FTDI Board

Mini-USB cable

LilyPad Protoboard

LilyPad LED

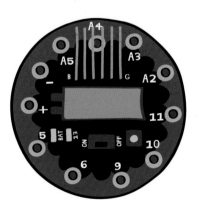

Lilypad Arduino SimpleSnap

LilyPad Speaker

Conductive Thread

CRAFT MATERIALS & TOOLS:

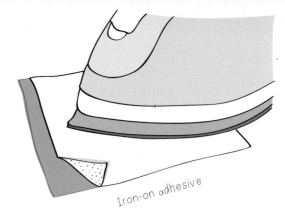

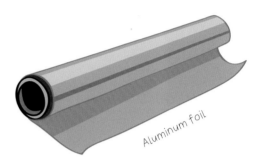

Iron-on adhesive

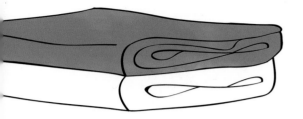

Fleece or felt for monster's body & decoration

Large-eyed Needle

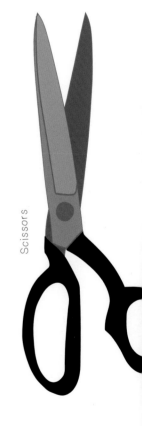

Aluminum foil

Paper

Embroidery thread or "floss"

Chalk or pencil for marking fabric

Scissors

Polyester filling

Colored pencils for drawing design sketches

Glue

GLUE

DESIGN YOUR MONSTER

YOUR ELECTRONICS

Before you start working on your project, it's good to know a bit more about the electronics you'll be using.

You're familiar with the LilyPad Arduino SimpleSnap, which was described in the last chapter. You've also worked with LEDs before, but the LilyPad SimpleSnap Protoboard (which we'll also call the Protoboard) and the LilyPad Speaker are new.

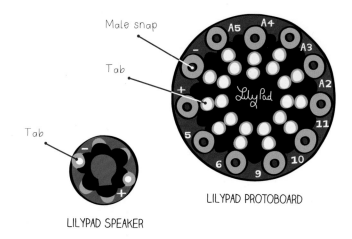

LILYPAD SPEAKER

LILYPAD PROTOBOARD

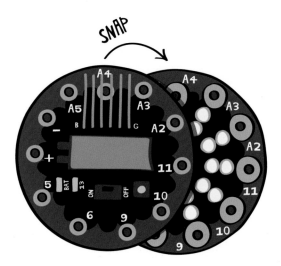

The Protoboard is the board with a ring of male snaps around its outer edge and holes in its center. The Protoboard snaps onto the LilyPad Arduino SimpleSnap. Try snapping the two boards together. Notice that they'll only snap together in one orientation. Try taking them apart—this can be tricky! If you're having trouble, wedge something stiff like a pair of scissors in between the two boards and gently pry them apart.

In this project you'll sew the Protoboard to your monster and snap the LilyPad to it. This will enable you to reuse your LilyPad in other projects. Notice how the Protosnap has labels on each of its snaps that correspond to the labels on the LilyPad.

The LilyPad Speaker, as you probably guessed, can make sounds when you send it the right kind of signal. It's also called the LilyPad Buzzer. You may see "Buzzer" instead of "Speaker" on your packaging. If so, don't worry—you still have the right part! This tutorial will call the board the LilyPad Speaker or simply the speaker.

BASIC DESIGN

Begin by designing a shape for your monster and drawing it on a piece of paper.

Your design should fit easily on a sheet of standard letter paper (8.5" x 11"). Cut out your shape to use as a template.

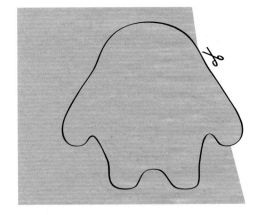

Note: As you design and build your monster you have to keep careful track of the front and back pieces and the inside and outside of your monster. All of the design drawings in this section show the outside of the monster.

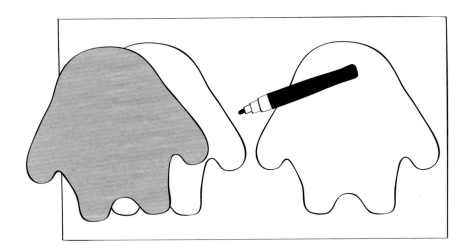

On a blank sheet of paper (ideally size 11"x 14"), use the template to trace out two copies of the monster, one next to the other.

Label one as the front of the monster and one as the back.

Decide on a color scheme for your monster and give it some personality. (You'll glue felt decorations to your monster after you build it.)

What will its eyes and mouth look like? Will it have claws? Toenails? Sketch these details on your design.

CIRCUIT DESIGN

Once you've created your monster character, decide where you want to put your Protoboard (the LilyPad will snap onto this board), LED, and speaker and add these components to your design.

Keep all of these pieces on the outside of your monster. Here, the Protoboard is on the back of the monster and the LED and speaker are on the front of the monster.

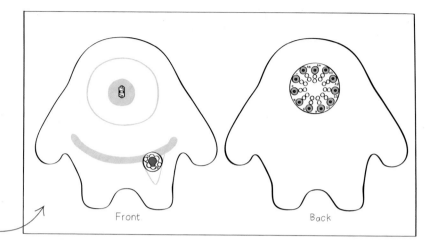

You can make your circuit diagram on your character sketch, or make a new drawing for the electrical elements. Do whatever you think will be most helpful.

Now that you know where the components will go, you need to figure out how to connect them to the Protoboard/Lily-Pad. Start with the LED.

The (-) tab on the LED should be attached to the (-) tab on the Protoboard. The (+) tab on the LED should be attached to one of the numbered tabs on the Protoboard—tab A4 for the monster here. (Remember that each of these connections is called a "trace" in the universe of electronics.) Sketch the (-) trace in black and the other trace in another color.

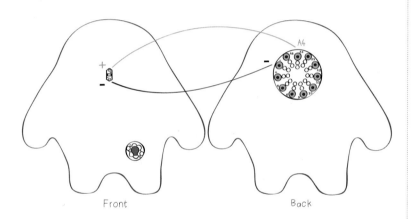

Next, plan the speaker connections. The (-) tab on the speaker also needs to be attached to the (-) tab on the Protoboard and the (+) tab on the speaker should be attached to another tab on the Protoboard—any tab except (-) or (+) will do. Here, the (+) tab of the speaker is attached to pin 5.

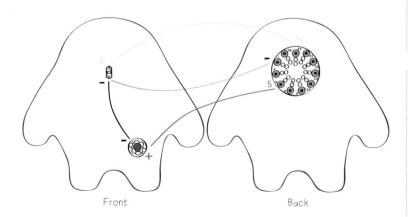

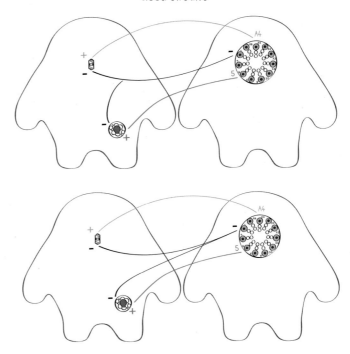

You may have noticed, in the drawing on the previous page, that the (-) side of the speaker is connected to the (-) side of the LED instead of the (-) tab on the Protoboard. You can connect the (-) tab of the speaker anywhere along the trace from (-) on the Protoboard to (-) on the LED. Electrically, all of these points are connected. It doesn't matter where you attach the (-) side of the speaker as long as it's somewhere along the (-) trace. The images on the right show other good ways to connect the speaker.

Use your colored pencils to add the connections to your drawing. Remember that black is used for (-) connections. The connections for your LED and speaker should be drawn in another color.

Bad circuits with crossing traces

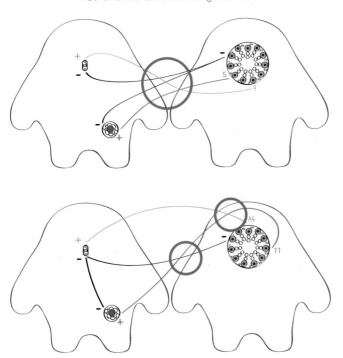

CIRCUIT PLANNING GUIDELINES

• Don't let traces from different pins or tabs cross each other or get too close to each other. When two different traces touch each other this creates a short circuit or "short" (like the ones described in the bookmark tutorial). In the monster, short circuits will cause the LED and speaker to malfunction and, in the worst case, may damage the LilyPad or its battery.

• It's OK for two of the same electrical traces to touch each other—like the (-) traces for the LED and speaker.

• Make sure you always plan out your circuit before you start building. Without careful planning, you will end up creating short circuits, sewing unnecessarily long traces, and having to take out and re-sew stitches during construction.

BEGIN BUILDING

CUT OUT YOUR FABRIC

Trace your template onto a your fabric
with a piece of chalk or a pencil. Make
two copies of the monster shape and
cut out both pieces—one for the front
and one for the back of the monster.

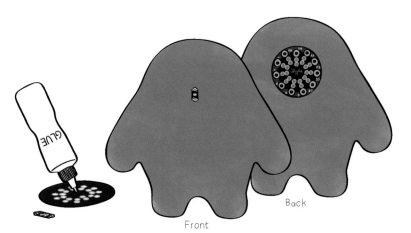

Front

Back

ATTACH YOUR COMPONENTS

Using your design sketch as a guide, glue
your Protoboard and LED onto the fabric
by putting a dab of glue on the backside
of each component. Use only as much
glue as you need to attach the pieces
and be careful not to fill any of their holes
since you'll need to stitch through them
soon.

SEW ONE SIDE TOGETHER

Using embroidery thread (or any other non-
conductive thread), sew one side of your
monster together. This will enable you to
stitch across the seam, from the front to the
back of the monster, with your conductive
thread.

Make sure that you keep all of your components on the side of the fabric that will
become the outside of your monster. Your
monster should look like the drawing on the
top of the next page when you're done.

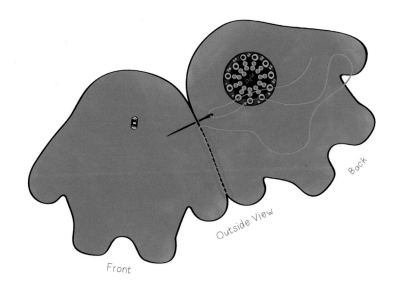

Front

Back

Outside View

DRAW YOUR CONNECTIONS

Using chalk or a pencil, draw the electrical connections between the LED and Protoboard. These will cross the seam you just sewed.

If you want to hide your stitches on the inside of your monster, draw these lines on the inside, where you'll be able to follow them. See the note at bottom of this page for more information about hiding your stitches.

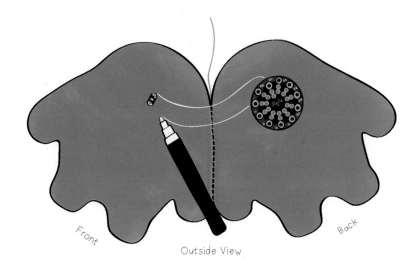

Front

Outside View

Back

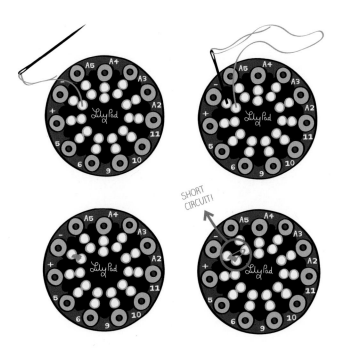

SHORT CIRCUIT!

SEW YOUR LED

Now you're ready to begin stitching the LED to the LilyPad Protoboard. Measure out 2-3' (1 meter) of conductive thread and thread your needle. Tie a knot on the underside of your fabric near the (-) holes on the Protoboard using the knot tying technique described on page 11 of the bookmark tutorial.

Sew tightly through the (-) holes at least three times to attach the Protoboard to the fabric and make a solid electrical connection between the Protoboard and your thread. (See the drawings on the left.)

Don't let the conductive thread touch any of the other holes on the Protoboard. This would create a short circuit.

Note: If you want, you can sew your conductive stitches only on the inside of the monster. This will mean that these stitches will not show up on the monster's outside. To hide your stitches, stitch only partway through the inside of your fabric with each stitch. This technique will only work with a thick fabric like a fleece or felt. The drawings in this tutorial show all conductive stitches sewn this way. However, you can also sew all the way through the fabric so that the stitches show up on both sides. Either stitching technique will work well.

Continue sewing from the Protoboard to the (-) side of the LED, following the connections that you drew on the fabric. You should use a running stitch, a simple stitch that is described on page 13, in the bookmark tutorial. Sew across the seam you just stitched together with embroidery thread. Note: The drawings on the right now show an inside (flipped over) view of the monster. Inside views are shown in light blue and outside views are shown in dark blue.

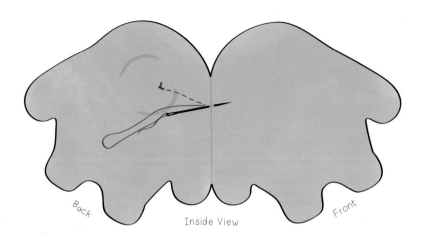

Inside View

Back

Front

When you reach the LED, sew at least three loops through the hole on the (-) side of the LED. Then, pull your thread tight and tie a knot on the inside of the monster. (You don't want an ugly knot on the outside!)

Cut the tails of the knots on both ends of your trace to about ¼" (6mm) long. Put a dab of glue on both of them to keep them from unravelling.

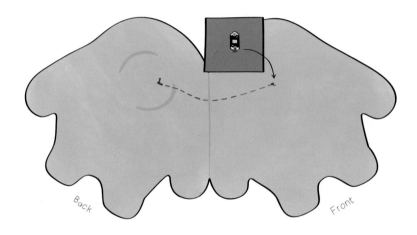

Back

Front

Now sew the (+) side of your LED to the correct pin on the Protoboard (A4, or the pin that you chose). Here's an inside view of the monster with both the (+) and (-) LED connections stitched in.

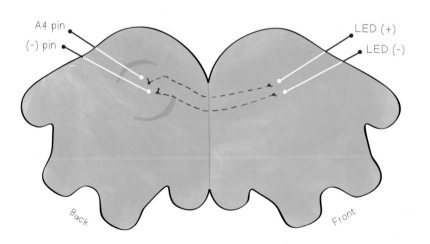

A4 pin

(-) pin

LED (+)

LED (-)

Back

Front

MAKE YOUR MONSTER BLINK

Snap the LilyPad onto the Protoboard on your monster. Turn on your computer, open up the Arduino software, and attach your LilyPad to your computer via the USB cable and the FTDI board.

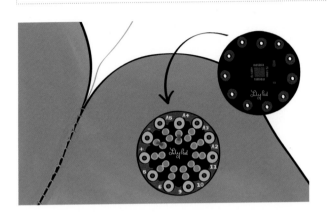

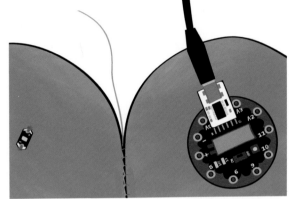

In the Arduino software, open up the "Blink" example (under File → Examples → 01.Basics → Blink). Delete all of the comments at the beginning of your program — all of the comments before the `int led = 13;` statement. This will make it easier to follow along with the examples here. Compile and upload this example to your LilyPad. A green LED on the LilyPad should begin to blink on and off.

You want to use this program to control the LED you've stitched into your monster, not the LED on the LilyPad. To do this you'll need to make a small edit. Notice that, at the beginning of the code, the variable `led` is set to the value 13 with the line `int led = 13;`. This tells the LilyPad which pin on the LilyPad the LED is attached to. When the variable `led` is set to 13, the code in the rest of the program will control the green LED that is on the LilyPad board. To get your monster's LED to blink, you should change 13 to the pin your LED is sewn to (A4 here). Make this change to the code and compile and upload it to your LilyPad. The LED you've sewn to your monster should begin to blink. If it doesn't, see the troubleshooting guide at the end of this section.

At this point, also delete unnecessary comments from the example code, as is shown on the right. Again, this will make it easier to follow along with this book's examples. Now you'll work through the code to understand what each line is doing.

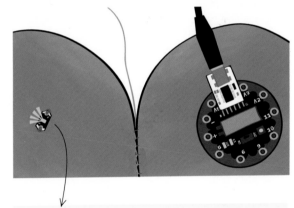

```
int led = A4;

// the setup routine runs once when you press reset:
void setup() {
    // initialize the digital pin as an output.
    pinMode(led, OUTPUT);
}

// the loop routine runs over and over again forever:
void loop() {
    digitalWrite(led, HIGH);  // turn the LED on
    delay(1000);              // wait for a second
    digitalWrite(led, LOW);   // turn the LED off
    delay(1000);              // wait for a second
}
```

PINMODE

There is one line in the setup section: pinMode(led, OUTPUT);. This line tells the LilyPad that the led pin (pin A4 for the program here) will be used to control an **output device** and should be in output mode. Outputs are things like lights, and motors that do things to the world. **Input devices,** which this tutorial will describe in a few pages, are things like switches and sensors that gather information about the world.

```
void setup() {
    pinMode(led, OUTPUT);
}
```

Whenever you add a component to your design you need to include a pinMode statement in the setup part of your program to tell the LilyPad whether the pin the component is attached to should be in input or output mode. (You'll do this shortly, when you connect your speaker to your Protoboard and begin programming it.)

pinMode is a **procedure** that is part of the Arduino programming language. The procedure pinMode has two **input variables** or "inputs": one specifies the pin that is being controlled and one specifies whether that pin will be an input or an output.

DIGITALWRITE

Now examine the statements in the loop section of your program. Remember from the programming tutorial that the loop section is where the main action of your program takes place.

```
void loop() {
    digitalWrite(led, HIGH);  // turn the LED on
    delay(1000);              // wait for a second
    digitalWrite(led, LOW);   // turn the LED off
    delay(1000);              // wait for a second
}
```

There are four statements in the loop section. The first one is:

```
digitalWrite(led, HIGH);  // turn the LED on
```

digitalWrite is another Arduino procedure. It is what turns your LED on and off. It takes two input variables, one that specifies which pin is being controlled and another that tells the LilyPad what to **write** or send to the pin. This statement sets a pin to either HIGH or LOW. This means that the pin, after this statement executes, will electrically be either HIGH or LOW.

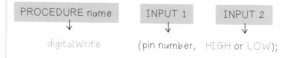

HIGH AND LOW

What do HIGH and LOW mean? These are code words that the Arduino language uses to talk about electricity. When the LilyPad encounters a digitalWrite(pin, HIGH); statement in the program, it sets the pin to (+). This statement tells the microcontroller on the LilyPad to close a microscopic switch that you can't see between (+) and the pin (remember the sparkling bracelet tutorial). When the LilyPad encounters a digitalWrite(pin, LOW); statement in the program, it sets the pin to (-)—the microcontroller closes a tiny switch between (-) and the pin. These digitalWrite statements are what allow you to control electrical signals with code. It's worth stopping to think about how powerful this is. The LilyPad and the Arduino turn text that you write on your computer into behavior that happens in the real world!

A summary of how the code that you write relates to real-world electricity is shown on the right. Note: The voltage of the LilyPad's built-in rechargeable battery is 3.7 volts. For more information, see the glossary entries for **high** and **low**.

CODE name	ELECTRICAL VALUE		
	symbol	voltage	other name
HIGH	+	3.7 volts	power
LOW	−	0 volts	ground

When you set the led pin (A4) HIGH with the statement digitalWrite(led, HIGH);, pin A4 gets set to (+). Since electricity flows from (+) to (-), current runs through the LED, lighting it up.

When you set the led pin (A4) LOW with the statement digitalWrite(led, LOW);, pin A4 gets set to (-). Since electricity does not flow from (-) to (-), the flow of electricity stops and the LED turns off.

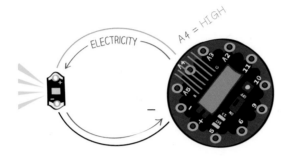

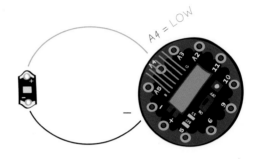

DELAY

delay is another procedure that you'll use all the time. It takes only one input—an amount of time in **milliseconds** (1/1000 second). The delay statement tells the LilyPad to stop and do nothing for the specified amount of time.

For example, the statement delay(1000); tells the LilyPad to do nothing for 1000 milliseconds, which is 1 second.

EXPERIMENT

Now that you're more familiar with the code, change the program to get a blinking behavior that you like for your monster.

Can you get your LED to flicker like a candle? Or thump like a heartbeat? You might also want to try using one of the LED behaviors you explored in the programming tutorial.

When you're finished, your loop section it should look something like the code on the right.

```
void loop() {
    digitalWrite(led, HIGH);
    delay(100);
    digitalWrite(led, LOW);
    delay(100);
    digitalWrite(led, HIGH);
    delay(500);
    digitalWrite(led, LOW);
    delay(500);
}
```

Your Blink code here

CREATE YOUR OWN PROCEDURE

So far you've used built-in procedures like digitalWrite and delay. Now you're going to write your own **procedure**. This new procedure will make your monster blink in the custom pattern you just created. A procedure is a chunk of code that is given its own special name. You're going to create a procedure called blinkPattern that will store the blink code you just wrote.

The diagram below shows the basic template for a procedure definition.

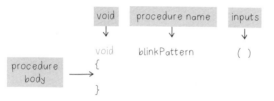

The definition begins with the word void followed by the name of the procedure. Any **inputs** taken by the procedure are listed in the parentheses to the right of the procedure's name. Since blinkPattern doesn't have any inputs, the parentheses are empty. Two curly brackets surround the "body"—the main part—of the procedure. These brackets tell the computer which statements are part of the procedure.

To create a procedure called blinkPattern, add the code below to the very end of your program, after the closing bracket of the loop section.

```
void blinkPattern()
{

}
```

Note: The name of the procedure doesn't matter to the compiler, so if you'd like to pick a different name for your procedure you can. Just make sure that you replace blinkPattern with the name of your procedure everywhere that it appears in the examples from here on out. Here, for example, is a procedure named myVerySuperSpecialBlinkyPattern:

```
void myVerySuperSpecialBlinkyPattern()
{

}
```

Try compiling and uploading the new code to make sure the basic procedure definition doesn't have any errors.

Next, copy the code from loop that defines your blink pattern and paste it into the body of your new blinkPattern procedure. Try compiling and uploading this code to make sure you haven't introduced any errors.

```
void blinkPattern()
{
    digitalWrite(led, HIGH);
    delay(100);
    digitalWrite(led, LOW);
    delay(100);
    digitalWrite(led, HIGH);
    delay(500);
    digitalWrite(led, LOW);
    delay(500);
}
```
Your Blink code here

Now, you need to use the name of the procedure, blinkPattern, somewhere else in the program to actually use the procedure. In programming slang this is known as **calling** a procedure. Edit your code so that it looks like the example below. Compile and upload this code. The behavior of your monster should be the same as it was before.

```
int led = A4;

void setup() {
    pinMode(led, OUTPUT);
}

void loop() {
    blinkPattern();
}

void blinkPattern()
{
    digitalWrite(led, HIGH);
    delay(100);
    digitalWrite(led, LOW);
    delay(100);
    digitalWrite(led, HIGH);
    delay(500);
    digitalWrite(led, LOW);
    delay(500);
}
```

Notice how you use the same format to call blinkPattern that you've used to call other procedures. Since blinkPattern doesn't have any inputs, the parentheses after its name are empty:

blinkPattern   ( );

You'll experience how useful your procedure is in a moment when your code starts to get more complex. For now, can you think of reasons why the ability to write procedures might be powerful? How might you use procedures to avoid repeating lines of code? Why might they make your programs easier to read? Why might they make your programs shorter?

SAVE YOUR CODE

Save your code by clicking on the downward pointing arrow in the Toolbar.

Click "OK" on the popup window that appears and choose a good name like "monster" for your file. Click on the "Save" button in the next window that appears to finish the process.

TROUBLESHOOTING CIRCUITRY AND CODE

GENERAL ADVICE

As you've probably already experienced, it can be tricky to identify and fix problems in projects that combine sewing and electronics. When you start to control your electronics with programs it becomes even harder. It's often difficult to tell from your project's behavior whether it's misbehaving because of an electrical issue, like a short, or a bug in your program, like a missing digitalWrite statement.

The troubleshooting pages in the rest of the book are separated into electrical diagrams that deal with circuitry and code diagrams that deal with programming. If your project isn't working properly and you don't know why, begin by following the electrical diagram, which will help you examine and fix your circuits. If this doesn't take care of the problem, work through the code diagram and look for programming mistakes.

You want to make sure you fix all electrical issues before you start worrying about programming glitches so you can tackle electrical and code problems separately. If there's a flaw in your circuitry, your project may not work even if your code is perfect. Make sure you've solved all of your electrical problems before you turn to code issues.

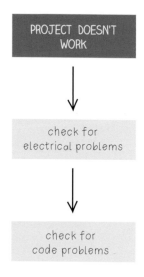

BASIC TECHNIQUES

Checking for and fixing electrical problems before you tackle programming errors is one important debugging strategy. Here are a few more:

- First and foremost, expect to have lots of errors! You'll almost certainly encounter several problems in every project. It's the nature of working with electronics and programs. Focus on solving the problems and try not to get frustrated.
- Always check your entire project for errors—all of your circuitry or all of your code. Don't assume you know where a problem is. It's often in a place where you don't expect it to be.
- Each time you make a change, check to see if it solved your problem. Don't make lots of changes before you check your project or you may introduce new errors.
- When you're looking for electrical issues, carefully follow each trace of stitching in your project. Look at the stitches on both the front and the back sides of the fabric. Look especially for loose connections around tabs and dangling threads that may be causing shorts.
- When you're looking for electrical issues, check to see what happens when you bend, jiggle, or squeeze the project. Changes in response to movement may indicate an electrical flaw like a loose connection or a short.
- When you're debugging your code, carefully read through your program from top to bottom, line by line. Take your time and don't skip anything. Try to visualize what each line is doing and compare what the code is saying to the behavior of your project. If you change anything in your code, recompile it and test it out right away.
- When you're printing information out on a computer screen (you'll do this in the next section), watch the values carefully. Check to see if they change in response to your behavior. Look for patterns in how they change. Printouts are a powerful troubleshooting tool.

TROUBLESHOOTING, LED

ELECTRICAL PROBLEMS

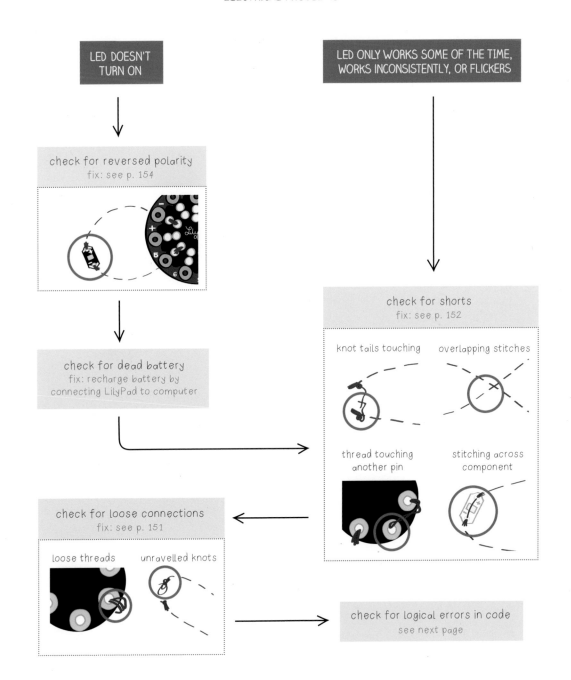

LED DOESN'T TURN ON

check for reversed polarity
fix: see p. 154

check for dead battery
fix: recharge battery by connecting LilyPad to computer

check for loose connections
fix: see p. 151

loose threads   unravelled knots

LED ONLY WORKS SOME OF THE TIME, WORKS INCONSISTENTLY, OR FLICKERS

check for shorts
fix: see p. 152

knot tails touching   overlapping stitches

thread touching another pin   stitching across component

check for logical errors in code
see next page

TROUBLESHOOTING, LED

CODE PROBLEMS

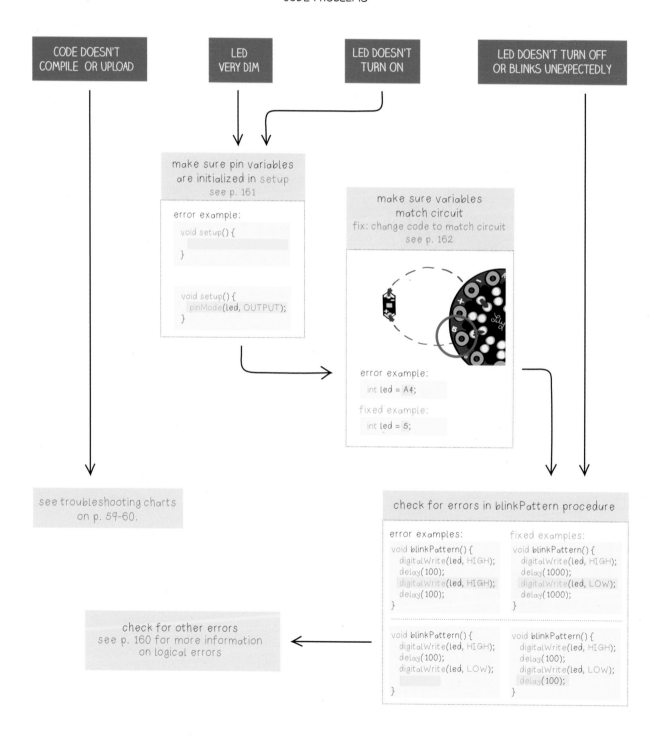

CODE DOESN'T COMPILE OR UPLOAD

LED VERY DIM

LED DOESN'T TURN ON

LED DOESN'T TURN OFF OR BLINKS UNEXPECTEDLY

make sure pin variables are initialized in setup
see p. 161

error example:

```
void setup() {

}
```

```
void setup() {
    pinMode(led, OUTPUT);
}
```

make sure variables match circuit
fix: change code to match circuit
see p. 162

error example:

```
int led = A4;
```

fixed example:

```
int led = 5;
```

see troubleshooting charts on p. 59-60.

check for errors in blinkPattern procedure

error examples:

```
void blinkPattern() {
    digitalWrite(led, HIGH);
    delay(100);
    digitalWrite(led, HIGH);
    delay(100);
}
```

fixed examples:

```
void blinkPattern() {
    digitalWrite(led, HIGH);
    delay(1000);
    digitalWrite(led, LOW);
    delay(1000);
}
```

```
void blinkPattern() {
    digitalWrite(led, HIGH);
    delay(100);
    digitalWrite(led, LOW);

}
```

```
void blinkPattern() {
    digitalWrite(led, HIGH);
    delay(100);
    digitalWrite(led, LOW);
    delay(100);
}
```

check for other errors
see p. 160 for more information on logical errors

ATTACH THE SPEAKER

Remove the LilyPad Arduino SimpleSnap from your Protoboard. Glue your speaker onto your monster. Remember not to fill its holes with glue.

Using your chalk or pencil, draw the electrical connections between the speaker and Protoboard. Draw these connections where you will be able to follow them with your stitching.

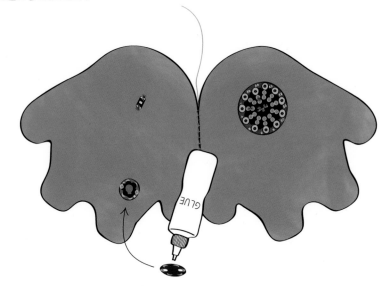

Sew the traces for your speaker. Stitch the (-) tab of the speaker to the (-) tab of the LED or anywhere along the (-) trace you've already sewn. Stitch the (+) tab of the speaker to the appropriate pin on the Protoboard. Here, the (+) side of the speaker is attached to pin 5. Once the speaker is sewn on, you can program it.

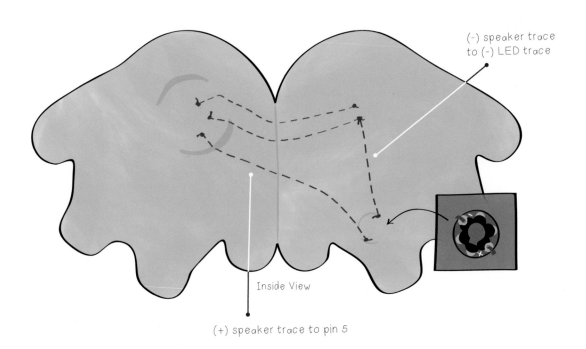

(-) speaker trace to (-) LED trace

Inside View

(+) speaker trace to pin 5

MAKE YOUR MONSTER SING

THE TONE PROCEDURE

You're going to use Arduino's built-in tone procedure to create sound. This is what a line of code using the tone procedure looks like:

```
tone(5, 1760);        // play a note
```

The two numbers in parentheses after tone are the inputs to the procedure. The first input is the pin the speaker's (+) tab is attached to and the second input is the frequency of the tone.

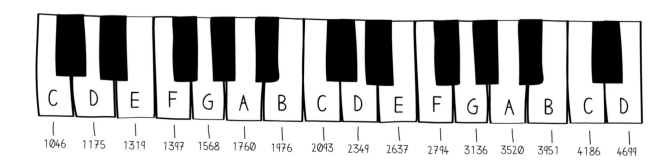

PROCEDURE name → tone
INPUT 1 → (pin number,
INPUT 2 → frequency);

• pin number: This input is the pin that the (+) end of your speaker is attached to. (Pin 5 in the example here.)

• frequency: This input is the **frequency** (or pitch) of the sound you want to play; the frequency is measured in Hertz (pulses per second). See the figure below for a chart of a few frequencies of familiar musical notes. (See http://www.sewelectric.org/MusicalNotes for a more extensive chart of the frequencies of different notes.)

C	D	E	F	G	A	B	C	D	E	F	G	A	B	C	D
1046	1175	1319	1397	1568	1760	1976	2093	2349	2637	2794	3136	3520	3951	4186	4699

Here is how you would use the tone procedure to play the note A, assuming your speaker's (+) tab is sewn to pin 5.

```
tone(5, 1760);
```

With this line, you're telling the tone procedure that your speaker is connected to pin 5 and you want to play a sound at a frequency of 1760 Hertz (note A).

A short overview of sound: Sound is created by vibrations of molecules in the air. When these vibrations hit your ear, your eardrums vibrate and you hear a note. When air molecules vibrate very quickly, you hear a high note; when they vibrate more slowly, you hear a low note. A speaker makes sound by vibrating and making the air vibrate at a particular speed, called a frequency. Inside the LilyPad speaker there is a material that moves in response to electricity. When this material moves it creates air vibrations and sound.

The tone procedure will play a note until you tell it to stop. The noTone procedure tells the LilyPad to stop playing a sound. The noTone procedure takes one input, the pin number that the (+) side of your speaker is attached to.

PROCEDURE name	INPUT 1
↓	↓
noTone	(pin number);

Here's how you would play the note A for one second (1000 milliseconds) and then pause for one second:

```
tone(5, 1760);    // note A begins playing
delay(1000);      // note plays for 1 second
noTone(5);        // note stops playing
delay(1000);      // silence for 1 second
```

MAKE A SOUND

Attach your LilyPad to your computer, snap it onto your Protoboard, and open the Arduino software.

Open the code you wrote in the last section by clicking on the upward pointing arrow on the Arduino Toolbar and selecting the file you saved earlier. When it's open it should look something like this:

```
int led = A4;

void setup() {
    pinMode(led, OUTPUT);
}

void loop() {
    blinkPattern();
}

void blinkPattern() {
    digitalWrite(led, HIGH);
    delay(100);
    digitalWrite(led, LOW);
    delay(100);
    digitalWrite(led, HIGH);
    delay(500);
    digitalWrite(led, LOW);
    delay(500);
}
```

Note: Since you are not going to edit the blinkPattern procedure now, the rest of this section does not include it in the program snapshots. However, you should still keep the blinkPattern procedure at the end of your code.

Add a variable called speaker to the code to tell the LilyPad what pin number your speaker is attached to.

This line should go right after the int led = A4; line. Since the (+) side of the speaker in the example here is attached to pin 5, the code here sets speaker to 5. You should set the speaker variable to the number of the pin your speaker is attached to.

```
int led = A4;
int speaker = 5;

void setup() {
    pinMode(led, OUTPUT);
}

void loop() {
    blinkPattern();
}
```

Next, set the pinMode for the speaker in the setup section. A speaker, like an LED, is an output—you're sending sound out into the world. So, you need to set the speaker pin to be an output. Add the line pinMode(speaker, OUTPUT); to your program, right after the line pinMode(led, OUTPUT);.

Also, so that you can focus on your sound generation code and not worry about your blink code, comment out the call to blinkPattern by adding // to the beginning of that line.

Upload this code to your LilyPad. Your LED should stop blinking, but your speaker won't produce any sounds yet.

```
int led = A4;
int speaker = 5;

void setup() {
    pinMode(led, OUTPUT);
    pinMode(speaker, OUTPUT);
}

void loop() {
    // blinkPattern();
}
```

```
int led = A4;
int speaker = 5;

void setup() {
    pinMode(led, OUTPUT);
    pinMode(speaker, OUTPUT);
}

void loop() {
    // blinkPattern();
    tone(speaker, 1760);
}
```

Now, add a call to the tone procedure to your loop section to get your speaker to play the note A.

Compile and upload this new code. You should hear a continuous tone from your speaker. If no sound comes out of your speaker, see the troubleshooting reference at the end of this section. If the sound starts to drive you crazy, unplug the USB cable from your computer and turn off your LilyPad.

To give yourself a break from the continuous beeping, add a noTone statement and some delays to your loop section, as shown below. Compile and upload the new code.

```
void loop() {
    // blinkPattern();
    tone(speaker, 1760);
    delay(500);
    noTone(speaker);
    delay(1000);
}
```

You may have noticed that it's not very intuitive to use numbers (frequencies) for musical notes. You can use variables to make working with tone easier and more intuitive.

Add a variable for one octave to the top of your program right before the setup section, as shown on the right.

```
int led = A4;
int speaker = 5;

int C = 1046;
int D = 1175;
int E = 1319;
int F = 1397;
int G = 1568;
int A = 1760;
int B = 1976;
int C1 = 2093;
```

Now you can use the names of notes in your code instead of their frequencies! You'll find that the notes are much easier to remember. Edit the loop section of your code, replacing frequencies with notes like the example shown on the right (top).

Now, experiment with the `tone` procedure. To change the note that it plays, call the procedure with a different note input. To change how long the note plays, adjust the delay after `tone`. For instance, the code shown on the right (bottom) shows how you would play the note C for one second.

Compile and upload the code each time you change it to try out your edits. Save your code by clicking on the downward pointing arrow in the Toolbar.

```
void loop() {
    // blinkPattern();
    tone(speaker, A);
    delay(500);
    noTone(speaker);
    delay(1000);
}
```

```
void loop() {
    // blinkPattern();
    tone(speaker, C);
    delay(1000);
    noTone(speaker);
    delay(1000);
}
```

PLAY A SONG

You can make the monster sing the first two verses of a song called *Hot Cross Buns* by editing the loop section so that it looks like the code on the right.

Try making this change to your program and compiling and uploading the new code. Can you recognize the song?

Notice how the two verses consist of three notes followed by a pause. Also notice how the three-note-phrase is repeated twice.

Now you're going to shorten this repetitive code by writing a procedure.

```
void loop() {
    // blinkPattern();
    tone(speaker, E);
    delay(2000);
    tone(speaker, D);
    delay(2000);
    tone(speaker, C);
    delay(2000);
    noTone(speaker);
    delay(2000);

    tone(speaker, E);
    delay(2000);
    tone(speaker, D);
    delay(2000);
    tone(speaker, C);
    delay(2000);
    noTone(speaker);
    delay(2000);

    delay(5000);
}
```

Create a procedure called `song` to store *Hot Cross Buns*. Add the code shown on the right (top) to your program, just after the closing bracket of the loop section but before the `blinkPattern` procedure definition.

This creates the basic skeleton for your `song` procedure, but there's no code in its body yet—inside its curly brackets. To give the `song` procedure some behavior, add the first verse of *Hot Cross Buns* to its body, as is shown here on the right (bottom).

```
void song() {
}

void blinkPattern() {
    digitalWrite(led, HIGH);
    delay(100);
    digitalWrite(led, LOW);
    delay(100);
    digitalWrite(led, HIGH);
    delay(500);
    digitalWrite(led, LOW);
    delay(500);
}
```

```
void song() {
    tone(speaker, E);
    delay(2000);
    tone(speaker, D);
    delay(2000);
    tone(speaker, C);
    delay(2000);
    noTone(speaker);
    delay(2000);
}

void blinkPattern() {
    ...
}
```

Next, to make use of your new procedure, you need to call it in the loop section of your code. Edit your loop section so that it looks like this:

```
void loop() {
    // blinkPattern();
    song();
    song();
    delay(5000);
}
```

Try compiling and uploading your code with this new addition to make sure you haven't made any errors.

The behavior of your monster should be the same as it was before. Notice that you've replaced 16 lines of code with two! Two calls to the song procedure replaced 16 lines of code.

This example illustrates one way that procedures are powerful: They enable you to give a name to a piece of code that is repeated. Then, instead of typing out the repeated code over and over again, you can just call the procedure.

PROCEDURE INPUTS

Procedures are powerful in other ways too. What if you wanted to make your song play faster? Say, hold each note for one second (1000 milliseconds) instead of two seconds (2000 milliseconds)? You could write a new procedure called fastSong that looks like the example on the right. (Note: Don't make any changes to your code yet. Just read along for a moment.)

```
void fastSong() {
    tone(speaker, E);
    delay(1000);
    tone(speaker, D);
    delay(1000);
    tone(speaker, C);
    delay(1000);
    noTone(speaker);
    delay(1000);
}
```

What if you wanted to make the song even faster, holding each note for ½ a second (500 milliseconds) instead of one second? Then you'd need a third procedure. And if you wanted to go faster still, you'd need a fourth!

This seems pretty crazy—especially since the basic behavior, the three notes played in a row, stays essentially the same. All you want to do is adjust how quickly the notes are played.

You can use another powerful feature of code to capture this range of possible speeds in a single procedure. You can add an **input variable** to your procedure that controls how long the notes are played. Change your song procedure to take an input called duration and change each instance of delay(1000); to delay(duration);, as shown on the right.

```
void song(int duration) {
    tone(speaker, E);
    delay(duration);
    tone(speaker, D);
    delay(duration);
    tone(speaker, C);
    delay(duration);
    noTone(speaker);
    delay(duration);
}
```

PROCEDURE name	INPUT 1
↓	↓
song	(duration);

The (int duration) addition in the first line tells Arduino that the procedure song requires an input called duration that's an int (short for an integer, a whole number). When you call song you now need to include a number for duration. The basic format of a call to song is shown on the left. Here's a call with a duration of two seconds: song(2000);

The example on the right shows how you'd get your original slow song to play. Make these edits to your code and try compiling and uploading it. You should hear the same song as before. When song runs, every instance of duration is replaced by the number you put in parentheses after song. In the example on the right, every duration in the song procedure would be replaced by 2000.

```
void loop() {
    // blinkPattern();
    song(2000);
    song(2000);
    delay(5000);
}
```

Now edit the loop section to take advantage of the new duration input to song like the example code to the right.

Compile and upload your new version of the code. Does the song sound different? Can you hear the speed changes? Save your code by clicking on the downward pointing arrow in the Toolbar.

```
void loop() {
    // blinkPattern();
    song(2000);      // play the song slowly
    song(1000);      // play the song faster
    song(500);       // and even faster
    delay(5000);
}
```

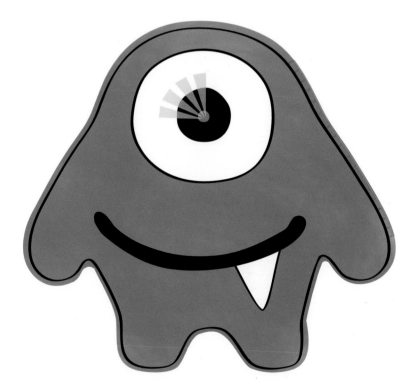

LAAAAAA
LALA
LAAAAA

EXPERIMENT

Try editing the body of the song procedure so that it plays a different tune. Can you get your monster to play *Mary Had a Little Lamb, Happy Birthday, Beethoven's Ninth Symphony,* or your own composition?

Hint 1: Below is a version of the song procedure that plays a scale:

Hint 2: You might want to play different notes for different amounts of time. There are a few different ways to approach this (each approach is described in a comment in the code below):

```
void song(int duration) {
    tone(speaker, C);
    delay(100);          // duration is permanently set at 100ms
    tone(speaker, D);
    delay(duration*2);   // duration is twice as long as input
    tone(speaker, E);
    delay(duration/2);   // duration is half as long as input
    noTone(speaker);
}
```

```
void song(int duration) {
    tone(speaker, C);
    delay(duration);
    tone(speaker, D);
    delay(duration);
    tone(speaker, E);
    delay(duration);
    tone(speaker, F);
    delay(duration);
    tone(speaker, G);
    delay(duration);
    tone(speaker, A);
    delay(duration);
    tone(speaker, B);
    delay(duration);
    tone(speaker, C1);
    delay(duration);
    noTone(speaker);
    delay(duration);
}
```

RECREATE A COMPLETE PROGRAM

Once you've created a melody that you like, you'll probably want to add the blinking behavior you created in the last section back to your program. Do this by removing the // characters that are in front of the blinkPattern procedure call in loop. Compile and upload your new code. See how your blinking and singing monster behaves. At this point, you may also want to adjust your blinkPattern or song procedures so that they work well together.

Notice how the song doesn't begin until your LED is finished blinking and, likewise, your blinking doesn't begin until the song has stopped. Remember that Arduino executes your program line by line in order. It can only do one thing at a time. It needs to finish one thing before it moves on to the next. In the example on the right, since blinkPattern(); is the first line in loop, the first thing the monster does is blink. Then, since song(2000); is the second line in loop, it plays its song. Then the monster delays (does nothing) for five seconds. Note: You may or may not want this last delay in your code. Eliminate it if you don't like the long pause.

```
void loop() {
    blinkPattern();
    song(2000);
    delay(5000);
}
```

SAVE YOUR CODE

Once you are happy with your program, save it by clicking on the downward pointing arrow in the Toolbar. Your entire program should now look something like the code on the right. The bodies of the song and blinkPattern procedures will be different in your code. The loop section may be slightly different as well.

```
int led = A4;
int speaker = 5; // speaker is attached to pin 5

int C = 1046;
int D = 1175;
int E = 1319;
int F = 1397;
int G = 1568;
int A = 1760;
int B = 1976;
int C1 = 2093;

void setup() {
    pinMode(led, OUTPUT);
    pinMode(speaker, OUTPUT);
}

void loop() {
    blinkPattern();
    song(2000);
    delay(5000);
}

void song(int duration) {
    tone(speaker, C);
    delay(duration);
    tone(speaker, D);
    delay(duration);
    tone(speaker, E);
    delay(duration);
    tone(speaker, F);
    delay(duration);
    tone(speaker, G);
    delay(duration);
    tone(speaker, A);
    delay(duration);
    tone(speaker, B);
    delay(duration);
    tone(speaker, C1);
    delay(duration);
    noTone(speaker);
    delay(duration);
}

void blinkPattern() {
    digitalWrite(led, HIGH);
    delay(100);
    digitalWrite(led, LOW);
    delay(100);
    digitalWrite(led, HIGH);
    delay(500);
    digitalWrite(led, LOW);
    delay(500);
}
```

TROUBLESHOOTING, SPEAKER

ELECTRICAL PROBLEMS

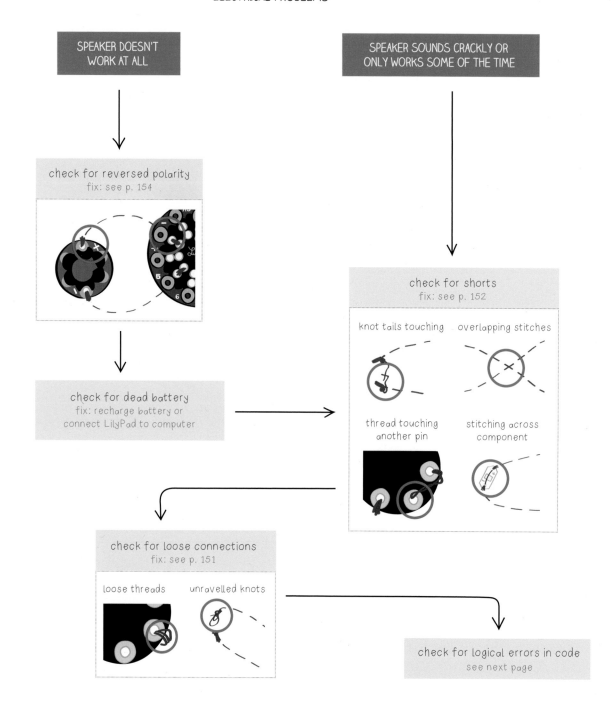

SPEAKER DOESN'T WORK AT ALL

check for reversed polarity
fix: see p. 154

check for dead battery
fix: recharge battery or connect LilyPad to computer

SPEAKER SOUNDS CRACKLY OR ONLY WORKS SOME OF THE TIME

check for shorts
fix: see p. 152

knot tails touching   overlapping stitches

thread touching another pin   stitching across component

check for loose connections
fix: see p. 151

loose threads   unravelled knots

check for logical errors in code
see next page

TROUBLESHOOTING, SPEAKER

CODE PROBLEMS

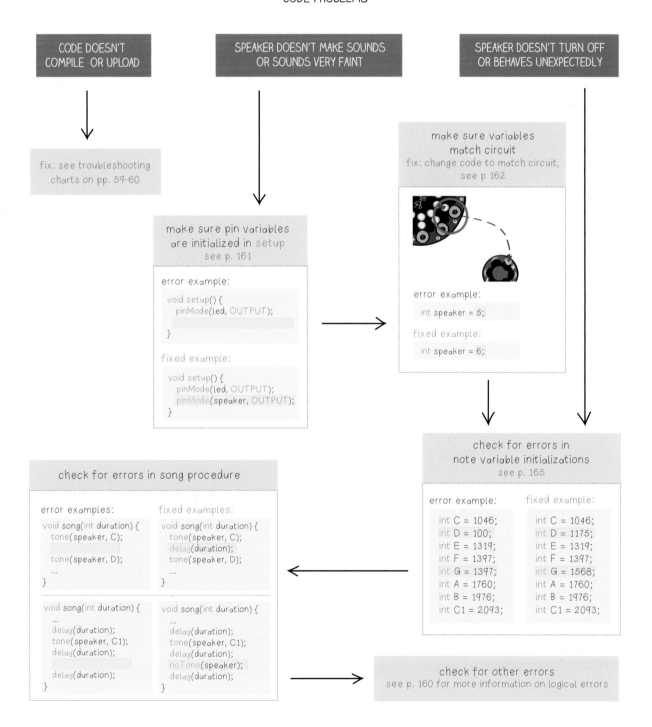

| CODE DOESN'T COMPILE OR UPLOAD | SPEAKER DOESN'T MAKE SOUNDS OR SOUNDS VERY FAINT | SPEAKER DOESN'T TURN OFF OR BEHAVES UNEXPECTEDLY |

fix: see troubleshooting charts on pp. 59-60

make sure pin variables are initialized in setup
see p. 161

error example:
```
void setup() {
    pinMode(led, OUTPUT);

}
```

fixed example:
```
void setup() {
    pinMode(led, OUTPUT);
    pinMode(speaker, OUTPUT);
}
```

make sure variables match circuit
fix: change code to match circuit, see p 162

error example:
```
int speaker = 5;
```

fixed example:
```
int speaker = 6;
```

check for errors in song procedure

error examples:
```
void song(int duration) {
    tone(speaker, C);

    tone(speaker, D);
    ...
}
```
```
void song(int duration) {
    ...
    delay(duration);
    tone(speaker, C1);
    delay(duration);

    delay(duration);
}
```

fixed examples:
```
void song(int duration) {
    tone(speaker, C);
    delay(duration);
    tone(speaker, D);
    ...
}
```
```
void song(int duration) {
    ...
    delay(duration);
    tone(speaker, C1);
    delay(duration);
    noTone(speaker);
    delay(duration);
}
```

check for errors in note variable initializations
see p. 165

error example:
```
int C = 1046;
int D = 100;
int E = 1319;
int F = 1397;
int G = 1397;
int A = 1760;
int B = 1976;
int C1 = 2093;
```

fixed example:
```
int C = 1046;
int D = 1175;
int E = 1319;
int F = 1397;
int G = 1568;
int A = 1760;
int B = 1976;
int C1 = 2093;
```

check for other errors
see p. 160 for more information on logical errors

GIVE YOUR MONSTER A SENSE OF TOUCH

This section describes how to add a touch **sensor** to your monster. Sensors allow projects to respond to inputs like switch presses or changes in light levels. In this case, a touch sensor will enable your monster to blink or play songs when someone holds its paws. The sensor you'll explore here is a **resistive sensor**.

DESIGN YOUR SENSOR

The sensor you're building consists of four aluminum foil patches on the monster's paws. These patches will detect when a person touches them. A person has to touch both paws at the same time for the sensor to "feel" the touch. The patches are attached to the monster's paws so that the monster can detect when someone holds its hands. Add these patches to your circuit design sketch.

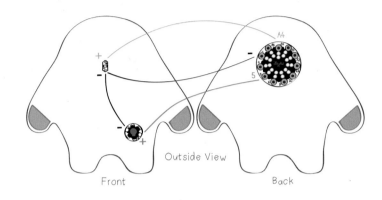

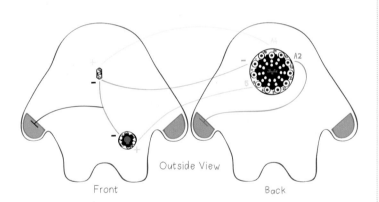

For the sensor to function electrically, one paw needs to be attached to (-) on the LilyPad and the other needs to be attached to tab A2, A3, A4, or A5. This paw *cannot* be attached to tab 5, 6, 9, 10, or 11. Here, the second paw is attached to tab A2. Though it looks like there are four sensors on the monster, note that when you stitch the monster together you'll only have two: one on each paw, covering both the front and back of the paws. Note: Make sure that when you put the monster together, your (-) trace and your A2 trace are not going to be sewn to different sides of the same paw!

Use colored pencils to add these details to your design sketch. Use black for the (-) paw and a new color for the A2 paw.

MAKE YOUR SENSOR

Get out your aluminum foil, iron-on adhesive, and iron. Read through the instructions on the heat-n-bond packaging to find the correct temperature setting for your iron and to familiarize yourself with the iron-on process. Place a piece of aluminum foil on your ironing board. Place a matching piece of iron-on adhesive on the aluminum foil. The rough and shiny (adhesive) side should be facing down. Iron the adhesive to the foil.

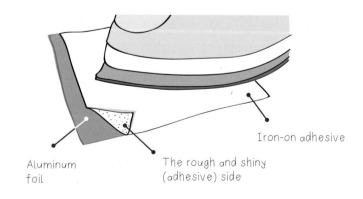

Make sure that the adhesive is firmly attached by peeling up a small corner of the paper. It should peel away easily and you should see a layer of clear adhesive stuck to the aluminum foil. Just peel up enough paper to check on the adhesive. Don't peel away the paper yet. Now you want to design and cut your sensors. On the paper that is attached to the aluminum foil, draw the shapes for your sensor patches. The monster template you made may be useful for this step.

Cut out your sensor patches, peel the paper off of them, and iron them onto your monster. Make sure that the adhesive side of the aluminum foil is facing your fabric. You don't want the glue to stick to your iron. It'll make a sticky mess!

SEW YOUR SENSOR

Next, mark the connections between your sensor patches and the rest of your circuit on your fabric using chalk or a pencil. Get ready to sew the connections between your patches and the rest of your circuit.

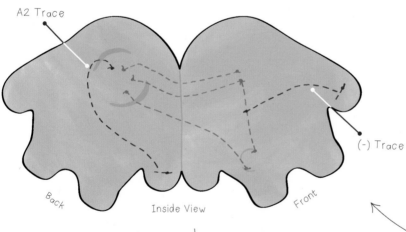

A2 Trace

(-) Trace

Back   Inside View   Front

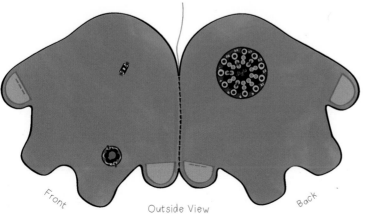

Front   Outside View   Back

On the back piece of the monster, sew from pin A2 (or the A-pin you chose) on the Protoboard to one patch. Make at least three running stitches through the aluminum foil to create a solid electrical connection between the patch and your conductive thread. Tie a knot and cut your thread.

On the front piece of the monster, sew from the (-) trace near your LED to the second sensor patch.

WRITE A PROGRAM TO DETECT PAW-TO-PAW CONTACT

Attach your LilyPad to your computer and snap it onto your Protoboard. Open the Arduino software. Open the code you wrote in the last section by clicking on the upward pointing arrow on the Arduino Toolbar and selecting the file you saved earlier.

Begin by adding a variable called aluminumFoil to the top of your program right below the int speaker = 5; line. This will tell the LilyPad which tab the aluminum foil patch is connected to.

You also need a pinMode statement like pinMode(led, OUTPUT); for your sensor to tell the LilyPad what kind of device is attached to the aluminumFoil pin. A sensor is an **input**, not an output, so the statement is slightly different; you use INPUT instead of OUTPUT.

```
int led = A4;
int speaker = 5;
int aluminumFoil = A2;

int C = 1046;
int D = 1175;
int E = 1319;
int F = 1397;
int G = 1568;
int A = 1760;
int B = 1976;
int C1 = 2093;

void setup() {
    pinMode(led, OUTPUT);
    pinMode(speaker, OUTPUT);
    pinMode(aluminumFoil, INPUT);
}
```

```
int led = A4;
int speaker = 5;
int aluminumFoil = A2;
int sensorValue;

int C = 1046;
int D = 1175;
int E = 1319;
int F = 1397;
int G = 1568;
int A = 1760;
int B = 1976;
int C1 = 2093;

void setup() {
    pinMode(led, OUTPUT);
    pinMode(speaker, OUTPUT);
    pinMode(aluminumFoil, INPUT);
    digitalWrite(aluminumFoil, HIGH);  // initializes the sensor
}

void loop() {
    blinkPattern();
    song(2000);
    delay(5000);
    sensorValue = digitalRead(aluminumFoil);
}
```

Now you're going to add several lines of code to your program. You're going to add a new variable called sensorValue, you're going to complete the initialization of the sensor pin in setup and you're going to read information from the sensor. The new lines are highlighted in the example program on the left. The tutorial will explore what each line does soon. For now, edit your code so that it looks like the example.

Try compiling and uploading this code. (Note: It won't do anything different than it did before yet.) If you encounter compile errors, read through your program carefully to make sure the code you added matches the example. Look especially for missing brackets "{" "}", and semicolons ";". If you continue to have problems, see the troubleshooting section.

DIGITALREAD

Now, look at the new line you've added to the loop section:

```
sensorValue = digitalRead(aluminumFoil);
```

This is the line that **reads** information from the sensor. Instead of **writing** information to a pin like digitalWrite, digitalRead reads information from a pin. When you want to gather information about actions happening in the world with an input device (like a sensor) you use a reading procedure like digitalRead. When you want to take action in the world with an output (like an LED or speaker) you use a writing procedure like digitalWrite.

PROCEDURE returns	PROCEDURE name	INPUT1
↓	↓	↓
HIGH or LOW	digitalRead	(pin number);

digitalRead is the first procedure you've encountered that **return**s a value. The basic procedure format is shown above. Returning means that the procedure gives something back to you when you call it. digitalRead tells you whether a pin is HIGH or LOW. It returns LOW when the pin is connected to ground (-) and HIGH when the pin is connected to power (+). A chart showing how these values relate to voltages is shown below.

value returned ↓	ELECTRICAL VALUE		
	symbol	voltage	other names
HIGH	(+)	3.7 volts	power
LOW	(-)	0 volts	ground

IF ELSE

Now that you know how to read information from your sensor you can use it to control your monster's behavior. Say you want your LED to light up if digitalRead returns LOW and turn off if digitalRead returns HIGH. The way you describe this situation in code is similar to the way you'd say it in a sentence. You use what's called a **conditional statement** to something like the statement on the right (top). Except, in your program you replace the word "otherwise" with the word "else".

The basic code structure of an if else conditional statement is shown on the right (middle). Notice how the body of the if and else statements are enclosed in curly brackets. Also notice how the **condition** is inside a pair of parentheses.

Now you'll write an actual conditional statement to see how the code's punctuation, structure, and behavior all come together. First, comment out everything in loop except the sensorValue = digitalRead(aluminumFoil); statement. (Note: You may have more than three lines that need to be commented out of your code. Comment out all of the blinking and song playing code before you move on to the next step.) Next, add an if else statement like the one shown on the right (bottom) to your loop section. Compile and upload your code.

When you touch the two sensor paws together, the paw connected to pin A2 and the paw connected to (-), the monster's LED should turn on. (Note: This may be a little awkward, but you need to make sure the paws that you stitched are touching. The paws with aluminum foil but no stitching won't work.) Save your program.

```
if digitalRead returns LOW
    the LED should turn on
otherwise
    the LED should turn off
```

```
if (condition)
{
    // do something
}
else
{
    // do something else
}
```

```
void loop() {
    // blinkPattern();
    // song(2000);
    // delay(5000);
    sensorValue = digitalRead(aluminumFoil);
    if (sensorValue==LOW)
    {
        digitalWrite(led, HIGH);
    }
    else
    {
        digitalWrite(led, LOW);
    }
}
```

WRITE A PROGRAM TO DETECT PERSON-TO-MONSTER CONTACT

The program you just wrote lets your monster change its behavior when you press its paws together. This is cool, but the goal is to have the monster respond when *you* touch it. To get the monster to respond to human touch, you'll need to use a different reading procedure. digitalRead can detect direct contact between the sensor pin (pin A2) and the ground pin (-). It can detect when the sensor pin is LOW (when its touching the ground pin) and when it's HIGH (when it's not touching the ground pin), but that's it. digitalRead can gather HIGH/LOW or **digital** information from switches—essentially, you close a switch when you touch the monster's two paws together.

A procedure called analogRead can gather more complex **analog** information from sensors. For the monster, analogRead can detect when the two paws are touching, and also when something slightly conductive (like a person) is touching the two paws, connecting them. Furthermore, as you'll see in a moment, it can even detect how hard you're pressing on the paws when you touch them!

To begin experimenting with analogRead, replace digitalRead(aluminumFoil); with analogRead(aluminumFoil); like the example shown on the right.

```
void loop() {
   // blinkPattern();
   // song(2000);
   // delay(5000);
   sensorValue = analogRead(aluminumFoil);
   if (sensorValue==LOW)
   {
      digitalWrite(led, HIGH);
   }
   else
   {
      digitalWrite(led, LOW);
   }
}
```

```
void setup() {
   pinMode(led, OUTPUT);
   pinMode(speaker, OUTPUT);
   pinMode(aluminumFoil, INPUT);
   digitalWrite(aluminumFoil, HIGH);
   Serial.begin(9600);              // initialize serial port
}

void loop() {
   // blinkPattern();
   // song(2000);
   // delay(5000);
   sensorValue = analogRead(aluminumFoil);
   Serial.println(sensorValue);    // send sensorValue to computer
   delay(100);                     // delay 1/10 of a second
   if (sensorValue==LOW)
   {
      digitalWrite(led, HIGH);
   }
   else
   {
      digitalWrite(led, LOW);
   }
}
```

Now you're going to add a few lines to your program that will let you see how analogRead works. You're going to send the readings taken by analogRead back to the computer, where you'll display them on your screen. You'll send the readings back through your **serial port**—the usb cable that connects your LilyPad to your computer.

First, add a line to initialize the serial port to the setup section. Then, add two lines to the loop section, as shown on the left. Compile and upload this new code.

Click on the magnifying glass icon in the upper-right-hand corner of the Arduino window. When you hover over this icon, a message that says "Serial Monitor" will appear.

A small window called the "Serial Monitor" will pop up. Any values your LilyPad's program sends back to the computer while it's running will appear in this window. You should see a steady stream of numbers. These values should be close to 1023.

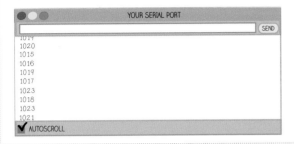

Now try touching your aluminum foil patches. Use one hand on each patch and make sure you're touching the two patches that are sewn to the Protoboard (see below). Watch the Serial Monitor. Try pressing your palms firmly against the pads and then lessening the pressure. The numbers should decrease noticeably to values in the 900s or below.

If you don't see any changes, make sure you're touching one hand firmly to each of the sewn patches of your sensor like in the diagram shown below. If your sensor still doesn't work—if the numbers in the Serial Monitor don't change—see the troubleshooting section.

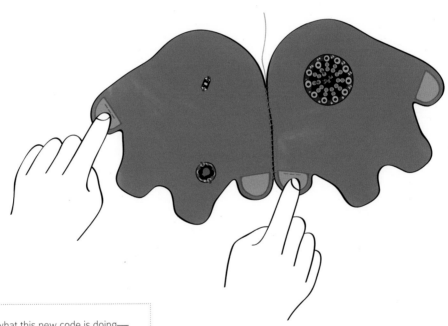

ANALOGREAD

Now you'll explore what this new code is doing— how it's translating your touch into numbers in the Serial Monitor.

Look at the heart of the code, the loop section. The first line after the commented-out code is: sensorValue = analogRead(aluminumFoil); Like digitalRead, this line **reads** information from your aluminum foil sensor. The information read from the sensor is then stored into the variable called sensorValue.

```
void loop() {
  // blinkPattern();
  // song(2000);
  // delay(5000);
  sensorValue = analogRead(aluminumFoil);
  Serial.println(sensorValue);
  delay(100);       // delay for 1/10 of a second
  ...
}
```

Like digitalRead, the analogRead procedure reads sensor data from a pin on the LilyPad. It takes one input, the pin number that your sensor is attached to, and gives back, or **returns,** a value that corresponds to the sensor reading. However, unlike digitalRead, which returns only two values, HIGH or LOW, analogRead returns a range of values, numbers between 0 and 1023. This means analogRead can give you more complex and nuanced information about what is happening in the world than digitalRead can.

The analogRead procedure works by measuring the voltage level on a pin. It returns a number between 0 and 1023 that corresponds to this voltage. The table on the right (bottom) shows how a few of the numbers it returns correspond to voltages.

PROCEDURE returns	PROCEDURE name	INPUT1
a number between 0 and 1023	analogRead	(pin number);

Value returned	ELECTRICAL VALUE		
	symbol	voltage	other names
1023	(+)	3.7 volts	power, HIGH
512		1.85 volts	
0	(-)	0 volts	ground, LOW

Returning to the sensor and the numbers that you're seeing in the Serial Monitor, why do the readings from the sensor change when you touch the sensor? Or, to put it a different way, why do the voltages measured at pin A2 change when you touch the sensor?

In the setup section of your program you set the sensor's initial value to HIGH or 3.7 volts with this line:

```
digitalWrite(aluminumFoil, HIGH);
```

This is why you see very high numbers, numbers close to 1023 (or HIGH), in the Serial Monitor when you're not touching the sensor. When you're not touching the sensor, analogRead returns the sensor's default value (HIGH).

When you touch the aluminum foil patches, one of your hands is connected to the (-) pin (which is at 0 volts, or LOW) and one of your hands is connected to pin A2 (which is at 3.7 volts, or HIGH). Your body, which is slightly conductive, creates an electrical connection between (-) and the sensing pin A2. This has the effect of lowering the voltage on pin A2, bringing it closer to 0 volts. This is why the numbers in the Serial Monitor go down when you touch the sensor.

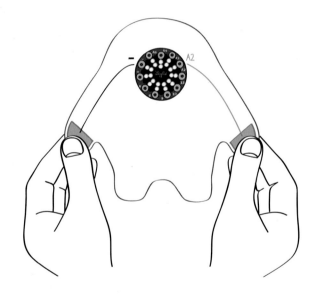

If you folded your monster in half and looked at it from the back, it would look like the drawing above.

The harder you squeeze on the sensor, the stronger the connection is between your body and the aluminum foil, and the better the electrical connection is between (-) and A2. This is why you see lower numbers as you squeeze harder. This is also why you see very low numbers (close to 0) when you touch the two paws directly together.

SERIAL.PRINTLN

The code you just wrote is doing another important thing that you haven't examined yet. It is sending information from the LilyPad back to the computer, where it is displayed in the Serial Monitor. The lines of your program that take care of this important communication function are in the setup and loop sections. Each line that is involved in the communication is highlighted on the right.

There are two important pieces to the communication code. First, in the setup section, you need to tell the LilyPad that it will be communicating with the computer and you need to tell the LilyPad how fast it should talk to the computer. The LilyPad and the computer can "talk" at different speeds and they need to be talking at the same speed to understand each other.

The line Serial.begin (9600); tells the LilyPad that it should talk at the speed of 9600 bits per second. You are using speed 9600 because it is the standard Arduino **communication speed**. This speed is also known as **baud rate** or **baud**.

```
void setup() {
  pinMode(led, OUTPUT);
  pinMode(speaker, OUTPUT);
  pinMode(aluminumFoil, INPUT);
  digitalWrite(aluminumFoil, HIGH);
  Serial.begin(9600);                // initialize serial port
}

void loop() {
  // blinkPattern();
  // song(2000);
  // delay(5000);
  sensorValue = analogRead(aluminumFoil);
  Serial.println(sensorValue);       // send sensorValue to computer
  delay(100);                        // delay 1/10 of a second
  if (sensorValue==LOW)
  {
    digitalWrite(led, HIGH);
  }
  else
  {
    digitalWrite(led, LOW);
  }
}
```

```
PROCEDURE name          INPUT
      ↓                    ↓
 Serial.begin      (communication speed);
```

The second piece of communication code is in the loop section. It's in loop that you actually send information from the LilyPad to the computer. In this program, you're sending the readings you took from your sensor back to the computer. The lines that do this are:

```
Serial.println(sensorValue);
delay(100);
```

The Serial.println procedure tells the LilyPad to send what is in its parentheses back to the computer, to be displayed in the Serial Monitor. Each value is printed on its own line. In the case of your code, the sensor readings that are read with analogRead and stored in the variable sensorValue, get sent back to the computer. These are the values you see in your Serial Monitor.

```
PROCEDURE name          INPUT
      ↓                    ↓
 Serial.println      (value to send);
```

There is one last important piece of the communication code, the statement delay(100);. This short delay gives the computer time to process and display the information that the LilyPad is sending. Without this delay, the computer will become overloaded with information from the LilyPad and crash. Whenever you have a Serial.println statement in your code, you also need a delay statement.

PUTTING IT ALL TOGETHER: CONTROL THE LED AND SPEAKER

Now you want to change your if else statement to take advantage of the numbers you're getting back from analogRead. You saw earlier, from the numbers in the Serial Monitor, that you get sensor values around 1023 when you aren't touching the monster, and that these values decrease when you squeeze the two paws.

Watch the serial monitor as you hold the monster's paws. Make a note of the highest number that you get when you are holding the paws. Write this number down on a sheet of paper. (Note: This number should be significantly lower than 1023. If it is not, you should see the troubleshooting section.)

Now you know that if the sensor values drop below the value that you wrote down, the monster's paws are being touched. You can think of this value as a touch-**threshold** value. If your sensorValue numbers are above the threshold, the monster isn't being touched. It they're below the threshold, the monster is being touched.

You can use this threshold information in your if else statement. Change your code so that the LED turns on if the monster is being touched and turns off if the monster is not being touched. To do this, you need to make one small change to your program.

```
void loop() {
    // blinkPattern();
    // song(2000);
    // delay(5000);
    sensorValue = analogRead(aluminumFoil);
    Serial.println(sensorValue);
    delay(100);
    if (sensorValue < 1000)
    {
        digitalWrite(led, HIGH);
    }
    else
    {
        digitalWrite(led, LOW);
    }
}
```

The code example on the left uses a threshold of 1000. Your threshold may be slightly lower than this, but it shouldn't be higher.

Make the change to your code and compile and upload the new version. What happens to your LED when you touch your sensors? What happens to the LED when you let go of the sensors?

If your monster isn't behaving predictably yet, adjust your threshold until the LED always turns on when you touch the sensors and always turns off when you let go.

Now, instead of turning the LED on and off when you touch your sensors, you can have it blink or sing, using the procedures you wrote earlier. Edit the code so that your monster sings when its paws are touched and blinks otherwise. Move the calls to blinkPattern and song into your if statement and remove the // characters on each of their lines. Your code should look something like the example shown on the right. Compile and upload your code. Save it and test out your monster.

Note: You may find that you have to hold your monster's hands for a long time before your song plays. This is because the LilyPad has to complete the blinkPattern procedure before it checks on the sensor. If your blinkPattern takes a long time to run, you'll have to wait a long time before your monster responds to your touch.

```
void loop() {
    sensorValue = analogRead(aluminumFoil);
    Serial.println(sensorValue);
    delay(100);
    if (sensorValue < 1000)
    {
        song(2000);
    }
    else
    {
        blinkPattern();
    }
}
```

EXPERIMENT

Experiment with your code to try out different behaviors. For example, can you have your monster play a song slowly when you're touching the paws lightly and quickly when you're touching the paws firmly?

SAVE YOUR CODE

When you're happy with your program, save it by clicking on the downward pointing arrow in the Toolbar.

Your entire program should now look something like the example on the right. Note that the bodies of the song and blinkPattern procedures will be different in your code and the loop section may be slightly different as well.

```
int led = A4;
int speaker = 5;
int aluminumFoil = A2;
int sensorValue;

int C = 1046;
int D = 1175;
int E = 1319;
int F = 1397;
int G = 1568;
int A = 1760;
int B = 1976;
int C1 = 2093;

void setup() {
    pinMode(led, OUTPUT);
    pinMode(speaker, OUTPUT);
    pinMode(aluminumFoil, INPUT);
    digitalWrite(aluminumFoil, HIGH);      // initializes the sensor
    Serial.begin(9600);                    // initializes the communication
}

void loop() {
    sensorValue = analogRead(aluminumFoil);
    Serial.println(sensorValue);
    delay(100);                            // delay for 1/10 of a second
    if (sensorValue < 1000)
    {
        song(2000);
    }
    else
    {
        blinkPattern();
    }
}

void song(int duration) {
    tone(speaker, C);
    delay(duration);
    tone(speaker, D);
    delay(duration);
    tone(speaker, E);
    delay(duration);
    tone(speaker, F);
    delay(duration);
    tone(speaker, G);
    delay(duration);
    tone(speaker, A);
    delay(duration);
    tone(speaker, B);
    delay(duration);
    tone(speaker, C1);
    delay(duration);
    noTone(speaker);
    delay(duration);
}

void blinkPattern() {
    digitalWrite(led, HIGH);
    delay(100);
    digitalWrite(led, LOW);
    delay(100);
    digitalWrite(led, HIGH);
    delay(500);
    digitalWrite(led, LOW);
    delay(500);
}
```

TROUBLESHOOTING, SENSOR

ELECTRICAL PROBLEMS

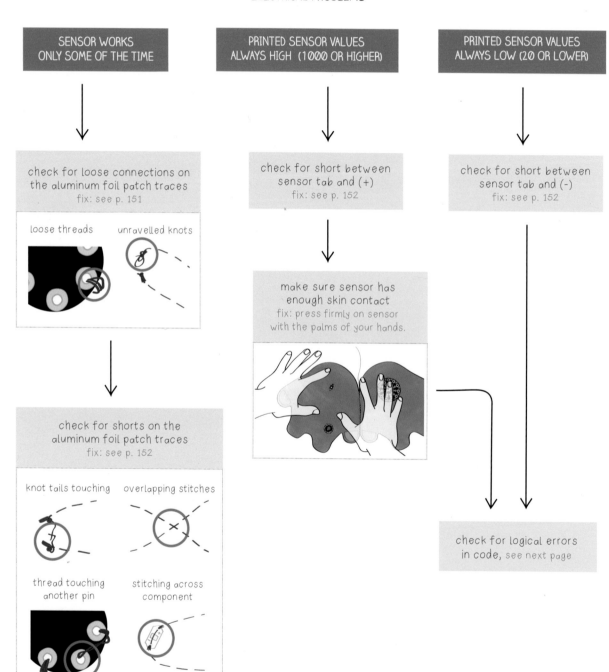

SENSOR WORKS ONLY SOME OF THE TIME

check for loose connections on the aluminum foil patch traces
fix: see p. 151

loose threads

unravelled knots

check for shorts on the aluminum foil patch traces
fix: see p. 152

knot tails touching

overlapping stitches

thread touching another pin

stitching across component

PRINTED SENSOR VALUES ALWAYS HIGH (1000 OR HIGHER)

check for short between sensor tab and (+)
fix: see p. 152

make sure sensor has enough skin contact
fix: press firmly on sensor with the palms of your hands.

PRINTED SENSOR VALUES ALWAYS LOW (20 OR LOWER)

check for short between sensor tab and (-)
fix: see p. 152

check for logical errors in code, see next page

TROUBLESHOOTING, SENSOR

CODE PROBLEMS

CODE DOESN'T COMPILE OR UPLOAD	PRINT OUT GARBLED OR NO PRINT OUT AT ALL	PRINTED SENSOR VALUES SEEM RANDOM	PRINTED SENSOR VALUES ALWAYS HIGH (1000 OR HIGHER)	PRINTED SENSOR VALUES CHANGE W/ TOUCH BUT NEVER TRIGGER LIGHT OR SOUND

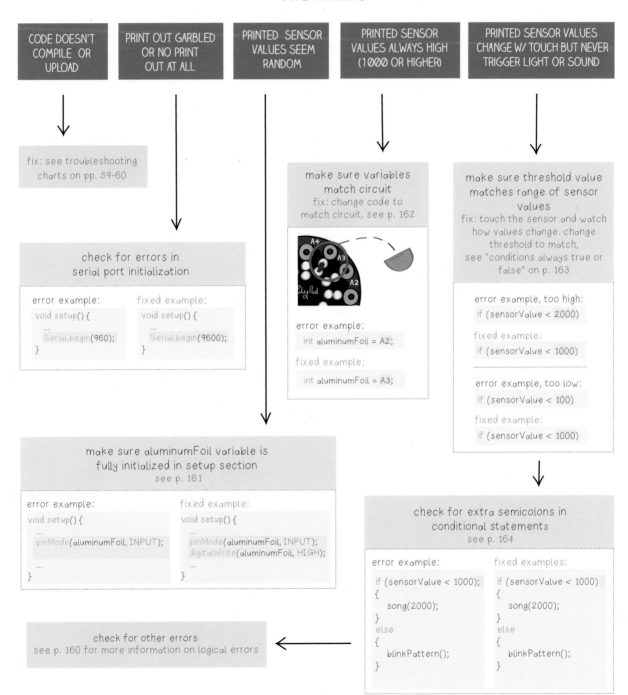

fix: see troubleshooting charts on pp. 59-60

check for errors in serial port initialization

```
error example:          fixed example:
void setup() {          void setup() {
   ...                     ...
   Serial.begin(960);      Serial.begin(9600);
}                       }
```

make sure variables match circuit
fix: change code to match circuit, see p. 162

```
error example:
   int aluminumFoil = A2;

fixed example:
   int aluminumFoil = A3;
```

make sure threshold value matches range of sensor values
fix: touch the sensor and watch how values change. change threshold to match, see "conditions always true or false" on p. 163

```
error example, too high:
   if (sensorValue < 2000)

fixed example:
   if (sensorValue < 1000)

error example, too low:
   if (sensorValue < 100)

fixed example:
   if (sensorValue < 1000)
```

make sure aluminumFoil variable is fully initialized in setup section
see p. 161

```
error example:                    fixed example:
void setup() {                    void setup() {
   ...                               ...
   pinMode(aluminumFoil, INPUT);     pinMode(aluminumFoil, INPUT);
                                     digitalWrite(aluminumFoil, HIGH);
   ...                               ...
}                                 }
```

check for extra semicolons in conditional statements
see p. 164

```
error example:              fixed examples:
if (sensorValue < 1000);     if (sensorValue < 1000)
{                           {
   song(2000);                 song(2000);
}                           }
else                        else
{                           {
   blinkPattern();             blinkPattern();
}                           }
```

check for other errors
see p. 160 for more information on logical errors

SEW AND STUFF YOUR MONSTER

Now that all of your monster's electronics are sewn on and tested, you can stitch the monster together!

Using the (non-conductive) embroidery thread, continue stitching the monster's two sides together along the outside edge of the monster until about 2" (5cm) of space remains open. Leave your embroidery thread uncut.

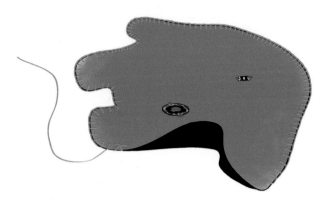

Stuff the inside of the monster with polyester filling until it's full. Stitch up the rest of the seam, so that your monster is completely enclosed. Tie off the embroidery thread, and snip off the ends of your knot.

Glue on the monster's eyes, mouth, claws, and other features.

GLUE

PLAY!

If your LilyPad is not attached to your monster, snap it on now. Unplug the Lily-Pad from your computer and turn it on.

Squeeze your monster, hold its hands, and listen to it sing! Take it into a dark room to see how bright and blinky it can be. Take it for a walk. Invite your friends over for a slumber party. Enjoy your new friend!

WASH

If your monster gets dirty, you can wash it by hand in cold water. Snap the LilyPad SimpleSnap off of the monster first to keep the battery dry!

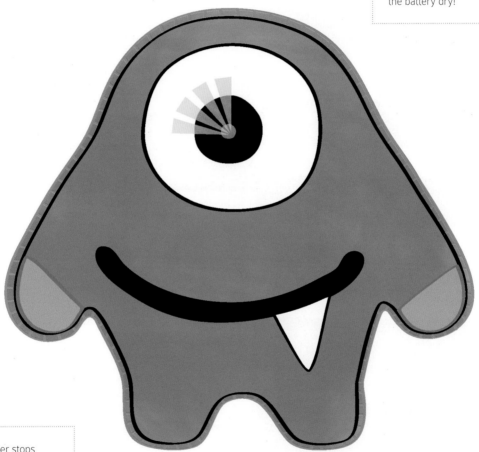

RECHARGE

If your monster stops working, its battery has probably died. Attach your LilyPad to your computer to recharge the battery.

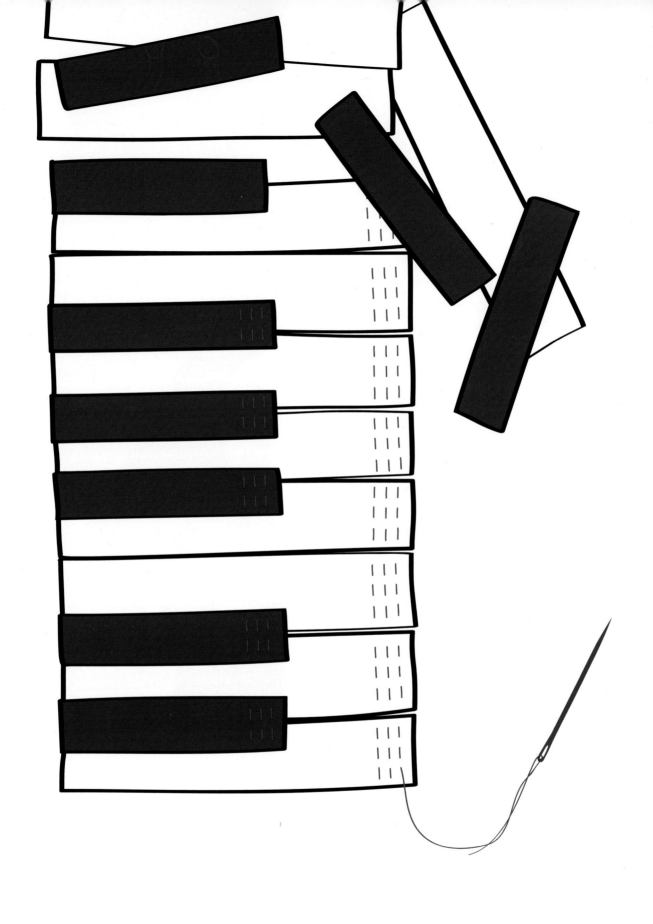

FABRIC PIANO

In the previous tutorial you learned how to program the LilyPad to make an interactive project. Now you'll explore how you can use the LilyPad to communicate with your computer. You'll build a soft piano that plays music both on your computer and through a sewn-in speaker.

Time required: 4-10 days

Chapter contents

Collect your tools and materials   p. 108
Design your piano   p. 110
Make a key   p. 112
Detect key presses   p. 114
Attach your speaker   p. 119
Make your piano sing   p. 120
Troubleshooting   *p. 123*

Finish building your piano   p. 126
Make your piano sing, part 2   p. 127
Troubleshooting   *p. 134*

Make readable printouts   p. 136
Condense your code with arrays   p. 139
Troubleshooting   *p. 146*

Make beautiful music   p. 147

COLLECT YOUR TOOLS AND MATERIALS

This project uses a LilyPad Arduino SimpleSnap, a speaker, and touch sensors made from conductive thread. You'll also need your sewing and sketching supplies.

ELECTRONIC MATERIALS:

Mini-USB cable

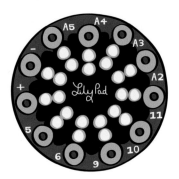

LilyPad Protoboard

LilyPad Speaker

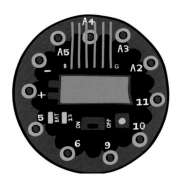

Lilypad Arduino SimpleSnap

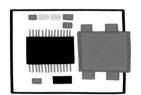

FTDI Board

Conductive Thread

CRAFT MATERIALS & TOOLS:

Fleece or Felt

Paper

Large-eyed Needle

Scissors

GLUE

Glue

Colored pencils

Chalk or pencil for marking fabric

DESIGN YOUR PIANO

PIANO KEYS AND CAPACITIVE SENSING

Before you begin designing, you should know a bit more about the "piano keys" that will be part of this project. Each key will be a touch sensor constructed out of conductive thread. The touch sensor you built for the interactive monster required you to touch two conductive patches—the monster's two paws. In this project, you'll use a different kind of touch sensor called a **capacitive sensor**. With a capacitive sensor, you only need to touch one conductive patch.

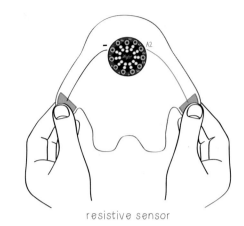

resistive sensor

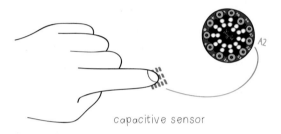

capacitive sensor

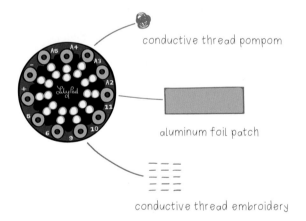

conductive thread pompom

aluminum foil patch

conductive thread embroidery

The circuitry for a capacitive sensor is simple. You attach a conductive patch or sew a conductive trace to a pin. That pin is then programmed to detect when a person touches the patch or trace. You'll explore how this touch sensing works in a moment. For now, it's important to know that the LilyPad will be able to detect when your skin contacts the conductive patch or trace. The larger the conductive area, the more sensitive the touch sensor will be. Three examples of sensors are shown on the left.

The piano you'll build has seven keys and a sewn-in LilyPad speaker. Each key will be a single trace of stitching in conductive thread connected to a pin on the LilyPad. The piano will be able to detect when you touch any one of these sensor keys.

BASIC DESIGN

Decide on a shape and color scheme for your piano and its seven keys. Draw out your pattern on a sheet of paper. Keep in mind that your sensor keys will need to be sewn to your LilyPad Arduino.

Be creative with your piano! The sensor keys don't have to be any particular shape or size.

sensor keys

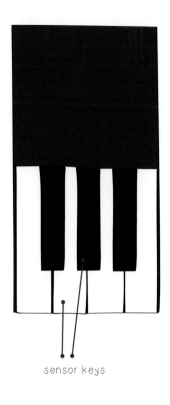

sensor keys

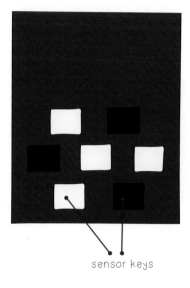

sensor keys

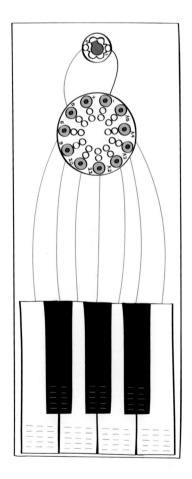

CIRCUIT DESIGN

Decide where you want to place your Protoboard and speaker. Draw out the trace for each sensor key. You can be creative with the conductive thread stitching for your sensors—you might use an embroidery stitch, spell something out in stitches, or sew out a decorative flower. Remember though to keep the traces for different keys as far apart as possible to prevent shorts.

The (+) tab on the speaker needs to be attached to one of the numbered tabs on the Protoboard and the (-) tab on the speaker needs to be attached to the (-) tab on the Protoboard. Draw these connections out on your sketch.

MAKE A KEY

PREPARE THE PIANO BASE

Trace your piano's shape onto your base fabric, and cut it out to create the base.

GLUE ON THE LILYPAD PROTOBOARD

Glue the LilyPad Protoboard onto the fabric.

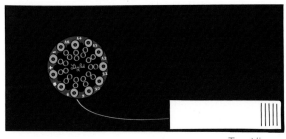

CREATE A PIANO KEY

Now you're ready to build your first piano key. Start with the far left key on your piano. In the design here, it is the key attached to pin 6.

First, create the decorations for the key you've designed (minus its conductive thread) and sew or glue these to your base fabric.

TRACE OUT YOUR CONNECTIONS

Using chalk or a pencil, draw out the traces for your key-sensors.

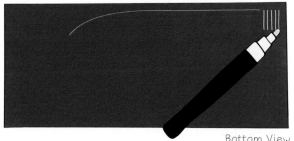

Top View

Bottom View

Identify the pin on your LilyPad protoboard that corresponds to your first key—pin 6 for the design here. Thread your needle with conductive thread. Tie a knot on the back of your fabric and make at least three loops through the appropriate pin holes on the Protoboard.

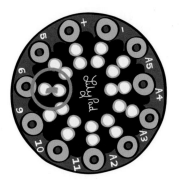

Stitch away from the protoboard to your key, following the path that you traced. Sew out any conductive key decorations you've designed. When you've finished your key, tie a knot on the back of your base fabric, trim its tails to ¼" (6mm) and seal them with a dab of glue.

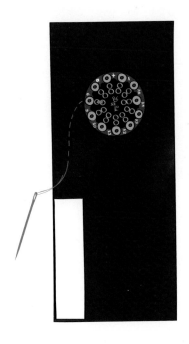

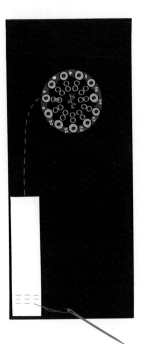

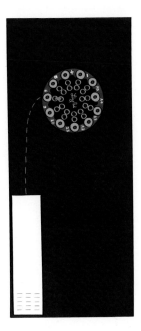

Bottom View

Stitching path shown in red

DETECT KEY PRESSES

Turn on your computer, open up the Arduino software, and attach your LilyPad Arduino SimpleSnap to your computer using the USB cable and FTDI board. Snap the LilyPad onto the Protoboard.

SEND DATA FROM THE LILYPAD TO THE COMPUTER

Open up a blank Arduino program by clicking on File → New. Add the basic setup and loop sections to this program, as shown below. Compile the program to make sure you haven't made any punctuation errors.

```
void setup() {

}

void loop() {

}
```

Now you'll write a program to send information from your LilyPad to your computer. Add one line to the setup section and two lines to the loop section, as shown below. Recall that the setup section runs once when the program starts, and then the loop section runs over and over until the LilyPad is turned off or reprogrammed.

```
void setup() {
   Serial.begin(9600);   // initialize the communication
}
void loop() {
   Serial.println("hey"); // send the word "hey" to the computer
   delay(100);            // delay for 1/10 of a second
}
```

The lines you've added should be familiar to you from the monster tutorial. The line Serial.begin(9600); initializes the communication between the LilyPad and the computer, telling the computer how fast the LilyPad is communicating (9600 bits per second, the standard Arduino communication speed).

The line Serial.println("hey"); sends the message "hey" to the computer. With the line delay(100);, you're pausing for 1/10 of a second so that you will not overload the computer by sending it too much data too quickly. Notice how comments here describe what each line does. You may want to add these comments or similar ones to your code for reference.

Upload this code to your LilyPad. If you encounter compile errors, read through your code carefully to make sure it matches the example. Look especially for missing or misplaced parentheses "(" ")", brackets "{" "}", and semicolons ";". Also remember that code is case-sensitive. If you are still having problems, see the troubleshooting chart on page 59 of the programming tutorial.

Once the code is successfully uploaded, click on the magnifying glass icon in the upper-right-hand corner of the Arduino window to open the Serial Monitor. You should see a steady stream of "hey"s.

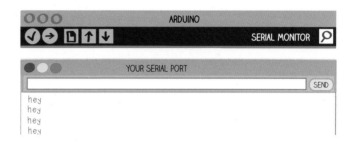

THE POWER OF OPEN SOURCE

Now you'll write the touch sensing part of your program to detect when your key is touched. Then you can send that information to the computer instead of "hey".

To detect whether a key is being touched or not, you'll use a procedure called readCapacitivePin. You're not going to write this procedure. Instead, you're going to use an **open source** procedure that was written by Mario Becker and Alan Chatham and posted online. Copy the code from this web page: http://www.sewelectric.org/CapacitiveSensingCode.

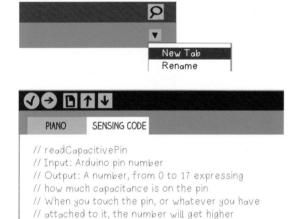

Create a new tab in your Arduino program to store this code. Click on the downward pointing arrow directly below the magnifying glass icon in the toolbar. On the menu that pops up, select "New Tab". You will be prompted to enter a name for the tab. Call it "sensingCode" or something that makes sense to you. Paste the code you've copied from the website into this tab.

Tabs help you keep long programs organized. Instead of keeping the code you just copied in your main program, you can keep it in its own tab. If you want to look at it or edit it you can click on that tab. Meanwhile, your main program stays simple and short.

The code you copied is long and complicated! But, don't worry. You don't have to understand the details of how it works to use it. It consists mainly of a **procedure** called readCapacitivePin. You've explored procedures in earlier tutorials. Here, you're encountering another one of the reasons that they are powerful and useful. You can share them! Moreover, you don't have to understand how one works to

use it. To use a procedure you only need to know four things about it:

1. Its *name*.

2. What it requires as *input*.

3. What it produces as *output*.

4. What it does.

structure of the
readCapacitivePin
procedure

OUTPUT	PROCEDURE name	INPUT
↓	↓	↓
number between 1 and 17	readCapacitivePin	(pin being measured);

The readCapacitivePin procedure you just downloaded works as follows:

1. *Procedure name*: The name is readCapacitivePin.

2. *The input*: The input is the number of the pin you want to turn into a sensor—the pin that your piano key is connected to.

3. *The output*: The procedure will return a number between 1 and 17 to indicate how hard the sensor is being touched. In your project, a 0 or 1 means that the key is not being touched and anything greater than 1 means the key is being touched.

4. *What it does*: This procedure uses an electrical property called capacitance to determine whether or not a person is touching a pin. (You'll learn more about capacitance in a few pages.)

Generally, procedures allow you to use a functionality without knowing how that functionality is implemented. In computer science, this idea is called **abstraction**—that is, you abstract away the details of *how* a procedure is written and instead focus on *what* it does. You can use the procedure without knowing exactly how it works.

If you think about it, you can find abstraction everywhere. People use lots of things without knowing how they work. You don't need to know what goes on inside a television in order to watch it; you don't need to know how to build a car in order to drive it; and you don't need to know how to make paper in order to sketch on it.

In the programming arena, abstraction allows people to write more complicated programs than they could write on their own. People can collaborate just by sharing code, without ever meeting one another. Now you're collaborating with Mario and Alan. It's code sharing and abstraction that let large groups of people cooperatively build massive and complex systems like the Internet.

Save your code by clicking on the downward pointing arrow in the Arduino software. Choose an appropriate name for your program like "Piano". Return to the main tab of your program—the tab labeled with the name of your program—by clicking on that tab.

WRITE A PROGRAM TO DETECT TOUCH

Now you'll explore the values that the readCapacitivePin procedure returns. To do that, you need to call the procedure and then send the values it returns to the computer where you can see them in the Serial Monitor. You'll make a few additions to your code to accomplish this task.

First, add two variables to your code. At the beginning of your program, create a variable called key1 and set its value to 6, (or the pin your first key is sewn to). This gives the name key1 to the pin that your first piano key is attached to. Create a second variable called touchValue. This variable will store the readings that you take from your sensor.

```
int key1 = 6;                    // name of the first sensor key
int touchValue;                  // will store sensor readings

void setup() {
    pinMode(key1, INPUT);    // set key1 to be an input
    Serial.begin(9600);          // initialize the communication
}

// the loop routine runs over and over again forever:
void loop() {
    Serial.println("hey");   // send the word "hey" to the computer
    delay(100);              // delay for 1/10 of a second

}
```

In the setup section of your code, set key1 to be an input. This tells the LilyPad that an input (in this case a touch sensor) will be attached to the key1 pin. Compile and upload this code to make sure you haven't introduced any errors.

Note: The behavior of the code won't change yet since you haven't added any new statements to the loop section.

Add the following line of code to the beginning of the loop section: touchValue = readCapacitivePin(key1);.
This new line makes use of the readCapacitivePin procedure. You're using key1 as the procedure's input—that's your touch sensing key—and you're storing the value returned by readCapacitivePin in the variable touchValue. Here's what the line looks like when broken down:

VARIABLE	ASSIGNED TO	PROCEDURE	INPUT PIN
↓	↓	↓	↓
touchValue	=	readCapacitivePin	(Key1);

```
int key1 = 6;                    // name of the first sensor key
int touchValue;                  // will store sensor readings

void setup() {
    pinMode(key1, INPUT);
    Serial.begin(9600);          // set key1 to be an input
                                 // initialize the communication
}

void loop() {
    touchValue = readCapacitivePin(key1); // read the touch sensor value
    Serial.println(touchValue);           // send touchValue to the computer
    delay(100);                           // delay for 1/10 of a second
}
```

You also want to send the value that you read from your sensor, touchValue, back to the computer. Change the Serial.println statement so that instead of sending "hey" back to the computer you send touchValue. Also, update all of the comments in your program so that they accurately describe what each line of code is doing.

Compile your code and upload it to your LilyPad. Open up your Serial Monitor. When you are not touching the key, a stream of 0s or 1s should appear in the Serial Monitor.

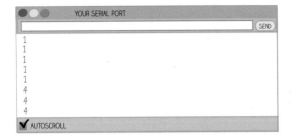

Now touch the key. Do the numbers change? If so, how?

The Serial Monitor on the left shows that the statement readCapacitivePin(key1); outputs the value 1 when no one is touching the key, and around 4 when someone is touching the key.

If your values are changing when you touch your key, you can move on to the next section. If they aren't changing, first carefully review your code to make sure it matches the example here. If you're still having problems, see the troubleshooting guide at the end of this section.

HOW DOES THE TOUCH SENSOR WORK?

Before you go on, it's worth exploring how the **capacitive sensor** you just made works.

As was mentioned before, it's different from the sensor you built in the monster tutorial. That touch sensor required two points of contact—you had to hold onto the monster with both hands, one hand on each conductive paw—for it to work. What's cool and useful about the sensor you just built is that it requires only one point of contact. You only need to touch one conductive area for it to trigger; you only need to use one hand.

The method of touch sensing you're using for the piano key is called capacitive sensing—a simplified version of the same method that a touch-screen phone or tablet uses to detect your gestures.

Capacitive sensing takes advantage of the electrical properties of the human body. All conductive materials (like conductive thread, aluminum foil, copper wire, and the human body) have **capacitance**, meaning that they can store electric charge. Imagine little bits of electric charge filling up a conductive area like the one below. Think of each circle as a bit of charge.

Each conductive material has a certain capacity to store electric charge, hence the name capacitance. When you touch a conductive material, you increase the amount of electric charge that it can store because the amount of material that is available to store charge increases—see the diagram below. When you touch a conductive material, its ability to store charge increases. In other words, its capacitance increases.

The readCapacitivePin procedure tells you how much capacitance the conductive material attached to a pin has. Since the capacitance changes when you touch the pin, it can detect your touch.

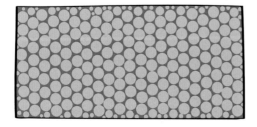

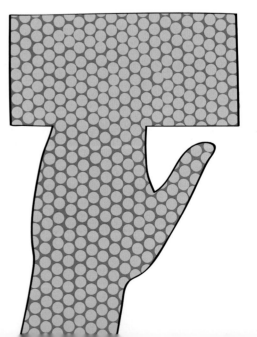

ATTACH YOUR SPEAKER

Now you're going to use your key presses to generate sound. The first step in that process is attaching your speaker.

SEW THE SPEAKER TO THE PIANO

Glue the speaker to your piano using your design sketch as a guide.

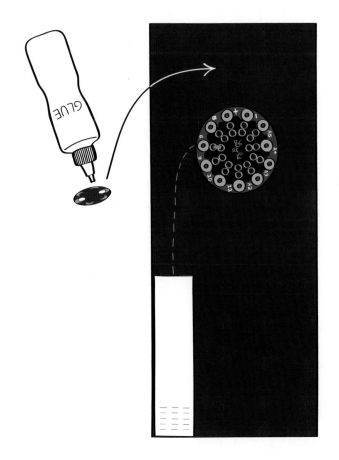

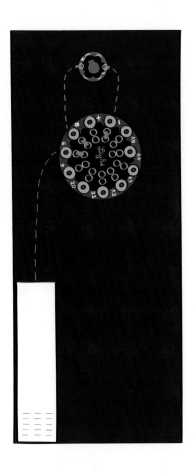

Sketch the speaker connections on your fabric with chalk or pencil. Here the speaker's (+) tab is attached to pin 5.

Begin sewing by tying a knot on the back of the fabric and looping through the (+) tab of the speaker. Then stitch from the (+) tab of the speaker to the appropriate tab on the Protoboard—tab 5 here. Once you've securely sewn the thread, tie another knot on the back of your fabric.

Next, sew the speaker's (-) tab. Begin by tying a knot on the back of the fabric near the speaker's (-) tab and stitching the (-) tab to the fabric. Sew from the speaker's (-) tab to the (-) tab on the Protoboard. After securely connecting to the (-) tab holes, tie another knot and cut your thread. Trim all of your knots and seal them with dabs of glue.

MAKE YOUR PIANO SING

Now you'll write code that will tell your LilyPad to play a note when you touch your key. Like you did in the monster lesson, you'll use a **conditional statement** to say:

• if a person is touching the key (that is, if you get a number greater than 1 back from readCapactivePin) the speaker should play a note.

```
if (condition)
{
    // do something
}
else
{
    // do something else
}
```

Recall from the monster tutorial that you can use the tone procedure to play a note and the noTone procedure to stop playing a note.

PROCEDURE name	INPUT 1	INPUT 2
↓	↓	↓
tone	(pin number,	frequency);

You'll also be using musical note frequencies again, so some of the frequency values for different notes are listed below. (These are reprinted from the monster tutorial.)

PROCEDURE name	INPUT 1
↓	↓
noTone	(pin number);

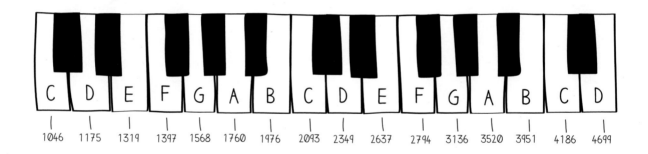

C	D	E	F	G	A	B	C	D	E	F	G	A	B	C	D
1046	1175	1319	1397	1568	1760	1976	2093	2349	2637	2794	3136	3520	3951	4186	4699

Now you'll use these elements to play some sounds. Begin by adding variables that map the frequencies in the piano diagram above to musical notes in your code. Add these variables right after the line int touchValue;

```
int touchValue;

// musical notes
int C = 1046;
int D = 1175;
int E = 1319;
int F = 1397;
int G = 1568;
int A = 1760;
int B = 1976;
int C1 = 2093;
int D1 = 2349;
```

Note: To preserve space, not all of the comments that were added to the code earlier are shown here, but you should keep all of the descriptive comments that you added to your code in place. They're shown again on the next page.

```
int key1 = 6;                       // name of the first sensor key
int speaker = 5;                    // name of the speaker key
int touchValue;                     // will store sensor readings

// musical notes
int C = 1046;
int D = 1175;
int E = 1319;
int F = 1397;
int G = 1568;
int A = 1760;
int B = 1976;
int C1 = 2093;
int D1 = 2349;

void setup() {
    pinMode(key1, INPUT);           // set key1 to be an input
    pinMode(speaker, OUTPUT);       // set speaker to be an output
    Serial.begin(9600);             // initialize the communication
}
```

Next, create and initialize a variable for your speaker. Create a variable called speaker and set its value to 5 (or the pin your speaker's (+) tab is sewn to). In the setup section add a line that sets this speaker pin to be an output.

Compile and upload this code. It won't do anything until you add some code to the loop section to actually make sound, but you want to make sure it compiles.

Edit the loop section of your code to play a note with your speaker. (Before you use your sensor key to trigger sound, you want to make sure the speaker actually works.) Use the tone and noTone procedures to play the note C on the speaker pin for one second. Upload this code to your LilyPad. Now your piano should be making noise. If it's not, see the troubleshooting guide.

```
void loop() {
    touchValue = readCapacitivePin(key1); // read the touch sensor value
    Serial.println(touchValue);            // send touchValue to computer
    delay(100);                            // delay for 1/10 of a second
    tone(speaker, C);                      // play the note C
    delay(1000);                           // wait for one second
    noTone(speaker);                       // stop playing the note
}
```

```
void loop() {
    touchValue = readCapacitivePin(key1); // read the touch sensor value
    Serial.println(touchValue);            // send touchValue to computer
    delay(100);                            // delay for 1/10 of a second
    if (touchValue > 1)                    // if the key is pressed
    {
        tone(speaker, C);                  // play the note C
        delay(100);                        // wait for 1/10th of a second
    }
    else                                   // if the key is not pressed
    {
        noTone(speaker);                   // stop playing the note
    }
}
```

You want your piano to play a note only when you touch your key. To do this you need to add a conditional statement that tells the piano to play a note only if the key is pressed.

Add an if else statement to your code so that the note C plays for 100 milliseconds if someone is touching the key and turns off otherwise. Note how the comments in the example on the left describe what each line of code is doing.

Compile and upload this code and test out your piano. It should now play a note each time you touch your key.

More specifically, whenever the key is touched, whenever touch-Value is greater than 1, the piano should play the note C and then wait for 1/10th of a second. The note will keep playing until the key is released, at which point the sound will cut off.

If your single-key piano is working well, save your code and go on to the next section. Your code should look more or less like the code on the right.

If your piano is not yet working, or if it is behaving erratically or unexpectedly, see the troubleshooting guide on the next page.

```
int key1 = 6;                                    // name of the first sensor key
int speaker = 5;                                 // name of the speaker key
int touchValue;                                  // will store sensor readings

// musical notes
int C = 1046;
int D = 1175;
int E = 1319;
int F = 1397;
int G = 1568;
int A = 1760;
int B = 1976;
int C1 = 2093;
int D1 = 2349;

void setup() {
    pinMode(key1, INPUT);                        // set key1 to be an input
    pinMode(speaker, OUTPUT);                    // set speaker to be an output
    Serial.begin(9600);                          // initialize the communication
}

void loop() {
    touchValue = readCapacitivePin(key1);        // read the touch sensor value
    Serial.println(touchValue);                  // send touchValue to computer
    delay(100);                                  // delay for 1/10 of a second
    if (touchValue > 1)                          // if the key is pressed
    {
        tone(speaker, C);                        // play the note C
        delay(100);                              // wait for 1/10th of a second
    }
    else                                         // if the key is not pressed
    {
        noTone(speaker);                         // stop playing the note
    }
}
```

TROUBLESHOOTING, ONE KEY

ELECTRICAL PROBLEMS

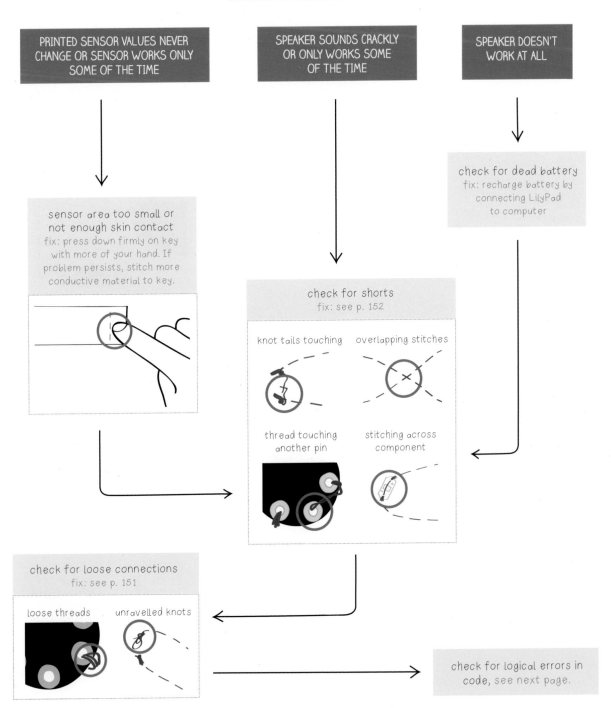

PRINTED SENSOR VALUES NEVER CHANGE OR SENSOR WORKS ONLY SOME OF THE TIME

SPEAKER SOUNDS CRACKLY OR ONLY WORKS SOME OF THE TIME

SPEAKER DOESN'T WORK AT ALL

check for dead battery
fix: recharge battery by connecting LilyPad to computer

sensor area too small or not enough skin contact
fix: press down firmly on key with more of your hand. If problem persists, stitch more conductive material to key.

check for shorts
fix: see p. 152

knot tails touching

overlapping stitches

thread touching another pin

stitching across component

check for loose connections
fix: see p. 151

loose threads

unravelled knots

check for logical errors in code, see next page.

TROUBLESHOOTING, ONE KEY

CODE PROBLEMS PART 1, SPEAKER

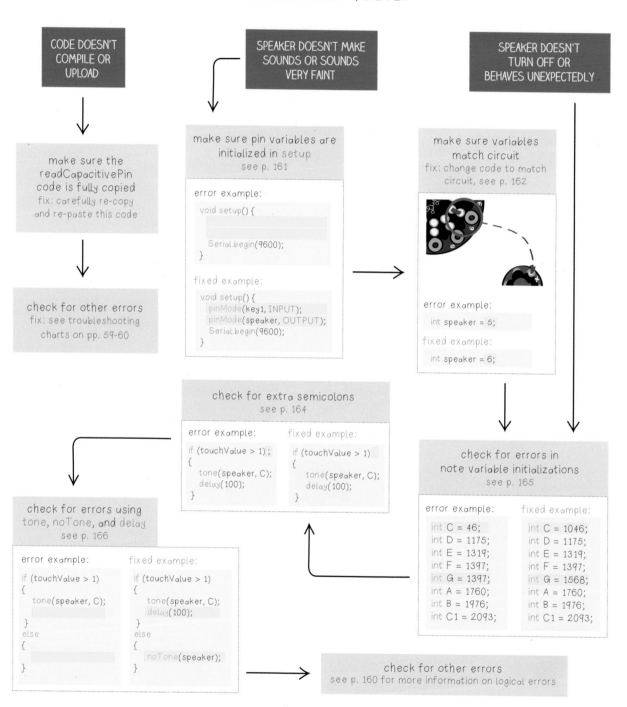

CODE DOESN'T COMPILE OR UPLOAD

make sure the readCapacitivePin code is fully copied
fix: carefully re-copy and re-paste this code

↓

check for other errors
fix: see troubleshooting charts on pp. 59-60

SPEAKER DOESN'T MAKE SOUNDS OR SOUNDS VERY FAINT

make sure pin variables are initialized in setup
see p. 161

error example:
```
void setup(){


    Serial.begin(9600);
}
```

fixed example:
```
void setup(){
    pinMode(key1, INPUT);
    pinMode(speaker, OUTPUT);
    Serial.begin(9600);
}
```

SPEAKER DOESN'T TURN OFF OR BEHAVES UNEXPECTEDLY

make sure variables match circuit
fix: change code to match circuit, see p. 162

error example:
```
int speaker = 5;
```

fixed example:
```
int speaker = 6;
```

check for extra semicolons
see p. 164

error example:
```
if (touchValue > 1);
{
    tone(speaker, C);
    delay(100);
}
```

fixed example:
```
if (touchValue > 1)
{
    tone(speaker, C);
    delay(100);
}
```

check for errors in note variable initializations
see p. 165

error example:
```
int C = 46;
int D = 1175;
int E = 1319;
int F = 1397;
int G = 1397;
int A = 1760;
int B = 1976;
int C1 = 2093;
```

fixed example:
```
int C = 1046;
int D = 1175;
int E = 1319;
int F = 1397;
int G = 1568;
int A = 1760;
int B = 1976;
int C1 = 2093;
```

check for errors using tone, noTone, and delay
see p. 166

error example:
```
if (touchValue > 1)
{
    tone(speaker, C);

}
else
{

}
```

fixed example:
```
if (touchValue > 1)
{
    tone(speaker, C);
    delay(100);
}
else
{
    noTone(speaker);
}
```

check for other errors
see p. 160 for more information on logical errors

TROUBLESHOOTING, ONE KEY

CODE PROBLEMS PART 2, SENSOR

PRINTOUT GARBLED OR NO PRINTOUT AT ALL

↓

check for errors in serial port initialization

error example:
```
void setup() {
  pinMode(key1, INPUT);
  pinMode(speaker, OUTPUT);
  Serial.begin(960);
}
```

fixed example:
```
void setup() {
  pinMode(key1, INPUT);
  pinMode(speaker, OUTPUT);
  Serial.begin(9600);
}
```

↓

make sure correct speed is selected in Serial Monitor
fix: select 9600 baud in lower-right corner of Serial Monitor window.

PRINTED SENSOR VALUES NEVER CHANGE

↓

make sure variables match circuit
fix: change code to match circuit, see p. 162

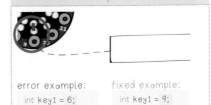

error example:
```
int key1 = 6;
```

fixed example:
```
int key1 = 9;
```

↓

check for errors using readCapacitivePin

error example:
```
readCapacitivePin(key1);
Serial.println(touchValue);
```

fixed example:
```
touchValue = readCapacitivePin(key1);
Serial.println(touchValue);
```

PRINTED SENSOR VALUES CHANGE W/ TOUCH BUT NEVER TRIGGER SOUND

↓

make sure threshold value matches range of sensor values
fix: touch the sensor and watch how values change. change threshold to match, see "conditions always true or false" on p. 163

error example, too high:
```
if (touchValue < 7)
```

fixed example:
```
if (touchValue < 1)
```

↓

check for errors in note variable initializations
see p. 165

error example:
```
int C = 46;
int D = 1175;
int E = 1319;
int F = 1397;
...
```

fixed example:
```
int C = 1046;
int D = 1175;
int E = 1319;
int F = 1397;
...
```

←

check for errors using tone, noTone, and delay
see p. 166

error example:
```
if (touchValue > 1)
{
    tone(speaker, C);

}
else
{

}
```

fixed example:
```
if (touchValue > 1)
{
    tone(speaker, C);
    delay(100);
}
else
{
    noTone(speaker)
}
```

→

check for other errors
see p. 160 for more information on logical errors

FINISH BUILDING YOUR PIANO

Now that you've got one key working, it's time to make the remaining six. First, build the decorations for your keys (minus their conductive thread) and attach these to your piano.

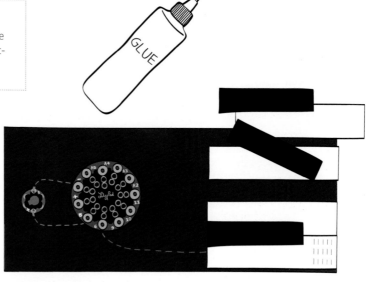

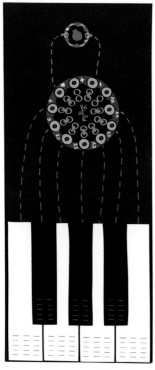

Top View

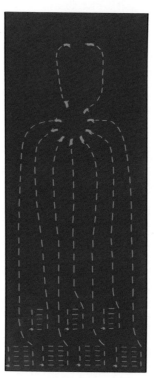

Bottom View

Stitch out the key-sensor traces. In the example here, the keys are attached to pins 6, 9, 10, 11, A2, A3, and A4.

For each key-sensor, begin by tying a knot on the back of the fabric near the appropriate pin on the Protoboard. Stitch snugly through the appropriate holes in the Protoboard (as always, at least three times) and then stitch from the Protoboard to your piano key. Once the key is complete, tie another knot on the back of your fabric. Trim your knots and seal them with glue before moving on to the next key.

Remember that each key must include exposed conductive thread since the exposed thread is what allows the LilyPad to detect touch.

MAKE YOUR PIANO SING, PART 2

```
// sensor keys
int key1 = 6;
int key2 = 9;
int key3 = 10;
int key4 = 11;
int key5 = A2;
int key6 = A3;
int key7 = A4;
int speaker = 5;          // name of the speaker key
int touchValue;           // will store sensor readings

//musical notes
int C = 1046;
int D = 1175;
int E = 1319;
int F = 1397;
int G = 1568;
int A = 1760;
int B = 1976;
int C1 = 2093;
int D1 = 2349;

void setup() {
    // set all keys to be inputs
    pinMode(key1, INPUT);
    pinMode(key2, INPUT);
    pinMode(key3, INPUT);
    pinMode(key4, INPUT);
    pinMode(key5, INPUT);
    pinMode(key6, INPUT);
    pinMode(key7, INPUT);
    pinMode(speaker, OUTPUT);   // set speaker to be an output
    Serial.begin(9600);         // initialize the communication
}
```

Now that you've built all of your keys, you need to program them. Open up the Arduino software and the Piano program you've been writing.

The first thing that you need to do is tell the program that you've now got seven keys to work with instead of one. You want to add some new ingredients to your recipe, some new key variables to your program.

Add variables for each of your new keys to the top of your program. Each of your keys is an input device (a sensor). You also need to initialize them in the setup section using pinMode statements. See the highlighted sections on the left.

Compile and upload this code to make sure you haven't introduced any errors.

Now you need to tackle the core of the program's behavior in the loop section. You'd like to program the LilyPad to play a different note for each key using the same code structure that you created for the first key. You need to write a conditional statement for each key that checks whether the key is being touched and plays a note if it is.

The body of the loop section currently looks like the code on the right. This is a basic block of code that detects whether a piano key is touched, sends that information to the computer, and plays a note if the key is touched.

You can use this same basic code block to control each piano key. You only need to change the key number and the note for the different keys. These variables are highlighted on the right.

```
touchValue = readCapacitivePin(key1);   // read the touch sensor value
Serial.println(touchValue);             // send touchValue to computer
delay(100);                             // delay for 1/10 of a second
if (touchValue > 1)                     // if the key is pressed
{
    tone(speaker, C);                   // play the note C
    delay(100);                         // wait for 1/10th of a second
}
else                                    // if the key is not pressed
{
    noTone(speaker);                    // stop playing the note
}
```

You could copy and paste this code into your program seven different times, one for each key. (Don't actually do this, just make a mental note that you could do things this way if you wanted to.) The code below shows what this would look like for the first three keys.

```
void loop() {
    touchValue = readCapacitivePin(key1);  // read the touch sensor value
    Serial.println(touchValue);            // send touchValue to computer
    delay(100);                            // delay for 1/10 of a second
    if (touchValue > 1)                    // if the key is pressed
    {
        tone(speaker, C);                  // play the note C
        delay(100);                        // wait for 1/10th of a second
    }
    else                                   // if the key is not pressed
    {
        noTone(speaker);                   // stop playing the note
    }

    touchValue = readCapacitivePin(key2);  // read the touch sensor value
    Serial.println(touchValue);            // send touchValue to computer
    delay(100);                            // delay for 1/10 of a second
    if (touchValue > 1)                    // if the key is pressed
    {
        tone(speaker, D);                  // play the note D
        delay(100);                        // wait for 1/10th of a second
    }
    else                                   // if the key is not pressed
    {
        noTone(speaker);                   // stop playing the note
    }

    touchValue = readCapacitivePin(key3);  // read the touch sensor value
    Serial.println(touchValue);            // send touchValue to computer
    delay(100);                            // delay for 1/10 of a second
    if (touchValue > 1)                    // if the key is pressed
    {
        tone(speaker, E);                  // play the note E
        delay(100);                        // wait for 1/10th of a second
    }
    else                                   // if the key is not pressed
    {
        noTone(speaker);                   // stop playing the note
    }

    ...

}
```

Notice how long and complicated this code is. And it's just for three keys! Think about all of the places where you might make an error—miss a semicolon or a bracket, say. Yikes!

Don't get too worried. Instead of creating this tangled and extensive program, you can leverage the power of **procedures** again. By writing a procedure for the basic code block, you can use the same code multiple times without rewriting it. You can **call** the procedure in loop instead of retyping (or copying and pasting) the same block of code over and over.

The basic format for defining a procedure is shown on the right (top). The definition begins with the word void. This is followed by the name of the procedure, in this case checkPianoKey. Inputs that the procedure requires are listed in the parentheses to the right of the procedure's name. Since checkPianoKey doesn't have any inputs yet, these parentheses are blank. Two curly brackets—an opening bracket at the beginning of the procedure and a closing bracket at the end—tell the computer where the procedure begins and ends.

Create a procedure called checkPianoKey and add it to the bottom of your program, immediately after the loop section.

```
void    procedure name    inputs
 ↓            ↓              ↓
void      checkPianoKey     ( )
{
                        procedure
}                         body
```

```
void loop() {
    ...
}

void checkPianoKey()
{
}
```

```
void checkPianoKey()
{
    touchValue = readCapacitivePin(key1);  // read the touch sensor value
    Serial.println(touchValue);            // send touchValue to computer
    delay(100);                            // delay for 1/10 of a second
    if (touchValue > 1)                    // if the key is pressed
    {
        tone(speaker, C);                  // play the note C
        delay(100);                        // wait for 1/10th of a second
    }
    else                                   // if the key is not pressed
    {
        noTone(speaker);                   // stop playing the note
    }
}
```

```
void checkPianoKey (int key)
{
    touchValue = readCapacitivePin(key);   // read the touch sensor value
    Serial.println(touchValue);            // send touchValue to computer
    delay(100);                            // delay for 1/10 of a second
    if (touchValue > 1)                    // if the key is pressed
    {
        tone(speaker, C);                  // play the note C
        delay(100);                        // wait for 1/10th of a second
    }
    else                                   // if the key is not pressed
    {
        noTone(speaker);                   // stop playing the note
    }
}
```

Begin writing the body of the procedure by copying and pasting the code that is currently in your loop section into checkPianoKey, so that your code looks like the example on the left. Make sure that the body code is between the two curly brackets. Compile the code to make sure it's error free.

You want the procedure to work for all of the keys, not just the first key, key1. You want to be able to tell the procedure which key it should check and play a note for. To do this, you need to add an **input variable** to your procedure. Call that variable key—a nice generic name. Add an input variable to the first line of your procedure. Remember that you need to specify the type of this input variable. In this case that type is int.

Next, change each occurrence of key1 in the body of the procedure to key so that the procedure will use the key input variable instead of always checking the first key, key1.

Adding an input variable to a procedure allows you to change the value of the variable in the procedure every time you use it. In this case, it allows you to call the procedure with a different key each time. As a result, you can use the same procedure to detect touch for every key.

To see how this works, edit the loop section of the program to make use of your new procedure. Because everything you were previously doing in the loop section is now in your procedure, you can replace all of the code in loop with one call to your procedure, using key1 as an input since that's the key you've been using so far. Notice how neat and simple your loop section is now!

```
void loop() {
    checkPianoKey(key1);
}
```

Note: Though it's not shown above, your checkPianoKey procedure should be at the bottom of your program, after the loop section. Meanwhile, the variable definitions and the setup sections are still in your program before the loop section.

Compile and upload this code to your LilyPad. Test it out. Your first key should work just as it did before. But, since you haven't added checks for any of your new keys they won't do anything yet.

Edit your code to add checks for each of your new keys. That is, call your checkPianoKey procedure once for each key, using that key's variable name as an input to the procedure like the code on the right.

Compile your code and upload it to your LilyPad. Try touching each of your keys. Does each one work? That is, does touching each key produce a sound? Does touching different keys produce different sounds? If your code looks like the code here, you've built a one-note piano.

```
void loop() {
    checkPianoKey(key1);
    checkPianoKey(key2);
    checkPianoKey(key3);
    checkPianoKey(key4);
    checkPianoKey(key5);
    checkPianoKey(key6);
    checkPianoKey(key7);
}
```

To fix the problem, first find where you produce sound in your code. The only place where you use the tone procedure to produce sound is in the checkPianoKey procedure. If you look at this line of code— tone(speaker, C);—you can see that it always produces the same sound; it always plays a C note. This is the heart of the problem.

You want the procedure to produce different sounds for different keys. You need to add a second input variable to the checkPianoKey procedure, one that specifies the note that you want the procedure to play. Call this second variable note. Change the procedure so that tone plays the variable note instead of always playing a C note.

```
void checkPianoKey (int key, int note)
{
    touchValue = readCapacitivePin(key); // read the touch sensor value
    Serial.println(touchValue);          // send touchValue to computer
    delay(100);                          // delay for 1/10 of a second
    if (touchValue > 1)                  // if the key is pressed
    {
        tone(speaker, note);             // play a note
        delay(100);                      // wait for 1/10th of a second
    }
    else                                 // if the key is not pressed
    {
        noTone(speaker);                 // stop playing the note
    }
}
```

Then, in the loop section of your code, you can include the notes that you want to play as part of your procedure calls. For each line, add a comma after your key name and add the name of the note you want that key to play.

Notice how adding input variables to a procedure allows you to make it much more versatile. Notice also how short and elegant your program is now. Compare it to the alarming tangle you would have had without the checkPianoKey procedure (shown on page 128). Procedures are amazing!

Compile your code, upload it to your LilyPad, and try out your keys. The code in the example on the right plays the notes of a pentatonic (five-note) scale. This scale has the nice property of sounding melodic no matter how you play.

```
void loop() {
    checkPianoKey(key1, C);
    checkPianoKey(key2, D);
    checkPianoKey(key3, E);
    checkPianoKey(key4, G);
    checkPianoKey(key5, A);
    checkPianoKey(key6, C1);
    checkPianoKey(key7, D1);
}
```

MAKE YOUR PIANO FASTER

Hold down one piano key continuously for awhile. What happens? Notice how the note turns on and then turns off before turning back on. Now play a bit with your piano. You may also notice that, in general, you have to hold your finger down on a key for a while before it plays a note—try playing your piano now and pay attention to the way you need to wait between key presses.

To make your piano more responsive, you can eliminate some of the delays in your program. Look at your check-PianoKey procedure. It has two delays, highlighted on the right.

```
void checkPianoKey (int key, int note)
{
    touchValue = readCapacitivePin(key); // read the touch sensor value
    Serial.println(touchValue);          // send touchValue to computer
    delay(100);                          // delay for 1/10 of a second
    if (touchValue > 1)                  // if the key is pressed
    {
        tone(speaker, note);             // play a note
        delay(100);                      // wait for 1/10th of a second
    }
    else                                 // if the key is not pressed
    {
        noTone(speaker);                 // stop playing the note
    }
}
```

The first delay was put in originally to make sure you didn't overload the computer with too much data from the Serial. println statement that precedes it. The second delay tells the piano to play a note for 1/10th of a second before moving on. This second delay is important. You want to make sure that a note plays for awhile after you press a key, but you might be able to shorten or eliminate the first delay.

You don't need to delay each time the checkPianoKey procedure is called. That is, you don't need to delay the program each time you check a key. However, you do need a delay somewhere to make sure that your computer doesn't get overloaded with information from the Serial.println statements.

```
void loop() {
   checkPianoKey(key1, C);
   checkPianoKey(key2, D);
   checkPianoKey(key3, E);
   checkPianoKey(key4, G);
   checkPianoKey(key5, A);
   checkPianoKey(key6, C1);
   checkPianoKey(key7, D1);
   delay(10);
}

void checkPianoKey(int key, int note) {
   touchValue = readCapacitivePin(key);   // read touch sensor value
   Serial.print(touchValue);               // send touchValue to the computer
   delay(100);                             // delay for 1/10 of a second
   if (touchValue > 1)                     // if the key is pressed
   {
      tone(speaker, note);                 // play a note
      delay(100);                          // wait for 1/10th of a second
   }
   else                                    // if the key is not pressed
   {
      noTone(speaker);                     // stop playing the note
   }
}
```

To shorten the total amount of delay while still making sure that you don't overload the computer, replace the first delay in the checkPianoKey procedure with a delay in the loop section. This way, instead of delaying for 700ms (7 x 100ms = 7/10ths of a second, almost one full second) each time loop runs, the program will only delay for 10ms (1/100th of a second).

Make this change to your code and compile and upload it to your LilyPad. Try playing your piano and notice how it's much more responsive than it was. If you hold down one key continuously, you'll notice that the note still turns on and off, but the delay between notes is much shorter.

Can you figure out why the note turns on and off even when you're holding the key down continuously? Note: Fixing this problem completely is somewhat challenging and this tutorial isn't going to describe the solution, but it's good to understand why things work the way they do. Hint: The problem is related to the delays in the program and the fact that noTone(speaker); turns all sounds off—not just the sound for a particular key.

You can also think of the behavior not as a problem, but as a feature of the instrument you've just built. The on/off behavior makes your piano sound interesting. Even if you could easily eliminate it, you might want to keep it. As the famous saying goes, "It's a feature not a bug!"

If your piano is behaving erratically or unexpectedly, see the troubleshooting section. Otherwise, save your program and go on to the next section. Your code should now look more or less like the code on the right.

```
// sensor keys
int key1 = 6;
int key2 = 9;
int key3 = 10;
int key4 = 11;
int key5 = A2;
int key6 = A3;
int key7 = A4;
int speaker = 5;                    // name of the speaker key
int touchValue;                     // will store sensor readings

//musical notes
int C = 1046;
int D = 1175;
int E = 1319;
int F = 1397;
int G = 1568;
int A = 1760;
int B = 1976;
int C1 = 2093;
int D1 = 2349;

void setup() {
    // set all keys to be inputs
    pinMode(key1, INPUT);
    pinMode(key2, INPUT);
    pinMode(key3, INPUT);
    pinMode(key4, INPUT);
    pinMode(key5, INPUT);
    pinMode(key6, INPUT);
    pinMode(key7, INPUT);
    pinMode(speaker, OUTPUT);        // set speaker to be an output
    Serial.begin(9600);             // initialize the communication
}

void loop() {
    checkPianoKey(key1, C);
    checkPianoKey(key2, D);
    checkPianoKey(key3, E);
    checkPianoKey(key4, G);
    checkPianoKey(key5, A);
    checkPianoKey(key6, C1);
    checkPianoKey(key7, D1);
    delay(10);
}

void checkPianoKey (int key, int note)
{
    touchValue = readCapacitivePin(key); // read the touch sensor value
    Serial.println(touchValue);            // send touchValue to computer
    if (touchValue > 1)                    // if the key is pressed
    {
        tone(speaker, note);               // play a note
        delay(100);                        // wait for 1/10th of a second
    }
    else                                   // if the key is not pressed
    {
        noTone(speaker);                   // stop playing the note
    }
}
```

TROUBLESHOOTING, ALL KEYS

ELECTRICAL PROBLEMS

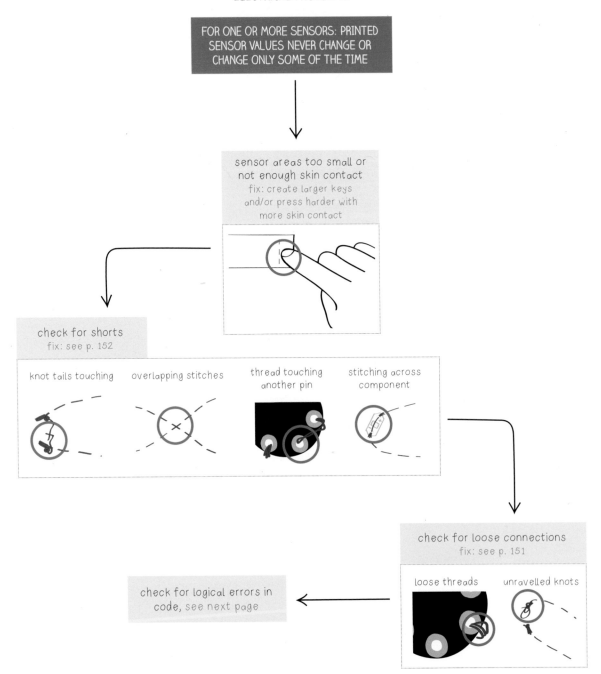

FOR ONE OR MORE SENSORS: PRINTED SENSOR VALUES NEVER CHANGE OR CHANGE ONLY SOME OF THE TIME

sensor areas too small or not enough skin contact
fix: create larger keys and/or press harder with more skin contact

check for shorts
fix: see p. 152

knot tails touching

overlapping stitches

thread touching another pin

stitching across component

check for loose connections
fix: see p. 151

loose threads

unravelled knots

check for logical errors in code, see next page

TROUBLESHOOTING, ALL KEYS

CODE PROBLEMS

CODE DOESN'T COMPILE OR UPLOAD

COMPUTER CRASHES WHEN SERIAL MONITOR IS OPENED

FOR ONE OR MORE SENSORS: PRINTED SENSOR VALUES NEVER CHANGE OR DON'T TRIGGER SOUND; SENSOR DOESN'T WORK

KEYS PLAY WRONG NOTES OR ALL PLAY THE SAME NOTE

fix: see troubleshooting charts on pp. 59-60

make sure variables match circuit
fix: change code to match circuit, see p. 162

check for errors in note variable initializations. see p. 165

error example:
```
int C = 46;
int D = 1175;
...
```

fixed example:
```
int C = 1046;
int D = 1175;
...
```

check for missing delay statement in loop

fixed example:
```
void loop () {
   checkPianoKey (key1, C);
   ...
   checkPianoKey (key7, D1);
   delay(100);
}
```

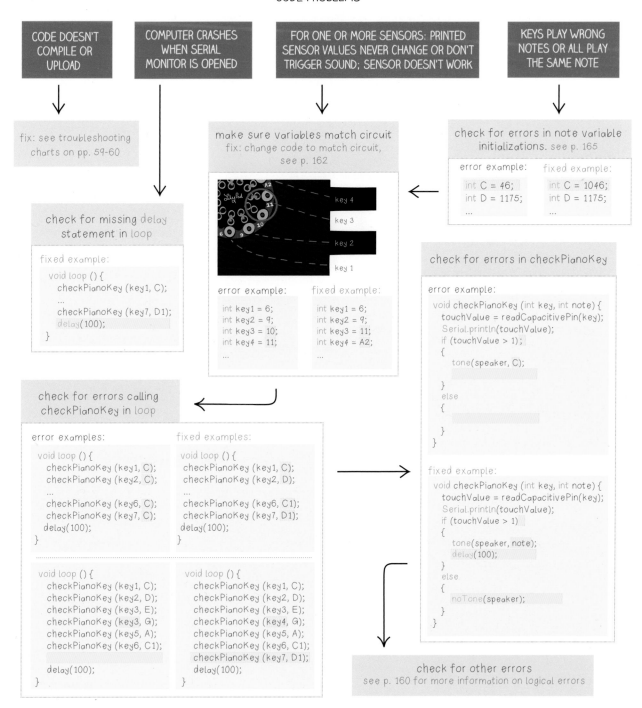

key 4

key 3

key 2

key 1

error example:
```
int key1 = 6;
int key2 = 9;
int key3 = 10;
int key4 = 11;
...
```

fixed example:
```
int key1 = 6;
int key2 = 9;
int key3 = 11;
int key4 = A2;
...
```

check for errors in checkPianoKey

error example:
```
void checkPianoKey (int key, int note) {
   touchValue = readCapacitivePin(key);
   Serial.println(touchValue);
   if (touchValue > 1);
   {
      tone(speaker, C);
   }
   else
   {

   }
}
```

fixed example:
```
void checkPianoKey (int key, int note) {
   touchValue = readCapacitivePin(key);
   Serial.println(touchValue);
   if (touchValue > 1)
   {
      tone(speaker, note);
      delay(100);
   }
   else
   {
      noTone(speaker);
   }
}
```

check for errors calling checkPianoKey in loop

error examples:
```
void loop () {
   checkPianoKey (key1, C);
   checkPianoKey (key2, C);
   ...
   checkPianoKey (key6, C);
   checkPianoKey (key7, C);
   delay(100);
}
```

fixed examples:
```
void loop () {
   checkPianoKey (key1, C);
   checkPianoKey (key2, D);
   ...
   checkPianoKey (key6, C1);
   checkPianoKey (key7, D1);
   delay(100);
}
```

```
void loop () {
   checkPianoKey (key1, C);
   checkPianoKey (key2, D);
   checkPianoKey (key3, E);
   checkPianoKey (key3, G);
   checkPianoKey (key5, A);
   checkPianoKey (key6, C1);

   delay(100);
}
```

```
void loop () {
   checkPianoKey (key1, C);
   checkPianoKey (key2, D);
   checkPianoKey (key3, E);
   checkPianoKey (key4, G);
   checkPianoKey (key5, A);
   checkPianoKey (key6, C1);
   checkPianoKey (key7, D1);
   delay(100);
}
```

check for other errors
see p. 160 for more information on logical errors

MAKE READABLE PRINTOUTS

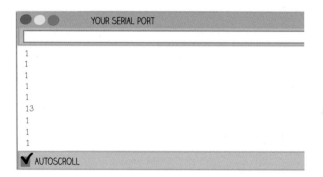

YOUR SERIAL PORT

```
1
1
1
1
1
13
1
1
1
```
✔ AUTOSCROLL

Begin this section by opening the Piano program you've been working on. If you haven't already, snap your Lily-Pad SimpleSnap onto your piano and plug it into your computer. Compile and upload your Piano program.

In the Arduino software, click on the magnifying glass in the Arduino window to open the Serial Monitor. You should see a stream of numbers that correspond to your sensor readings, like the ones on the left.

Try touching your keys and watch the numbers change. Notice how it's difficult to know which number corresponds to which sensor. This printout isn't formatted very well. Now you're going to learn how to format this printout to make it more legible. You can do this by printing the sensor values for all seven keys on a single line, so that the printout will look like the image on the right.

YOUR SERIAL PORT | SEND

17	1	17	1	1	1	1
17	1	17	1	1	1	1
17	1	17	1	1	1	1
17	1	17	1	1	1	1
17	1	17	1	1	1	1
17	1	17	1	1	1	1
17	1	17	1	1	1	1
17	1	17	1	1	1	1
17	1	17	1	1	1	1

✔ AUTOSCROLL

In the checkPianoKey procedure, where you do your printing, change the Serial.println statement to a Serial.print statement. Serial.println prints out touchValue followed by a carriage return. A carriage return creates a new line, like a hit to the return key on a keyboard. So, when you use Serial.println the next value you print shows up on the next line or ln. The new Serial.print statement—notice how the ln is missing from the end—prints out touchValue, but no carriage return (so the next value you print shows up on the same line, right after the last touchValue).

```
void checkPianoKey(int key, int note) {
    touchValue = readCapacitivePin(key); // read touch sensor value
    Serial.print(touchValue);              // send touchValue to the computer
    if (touchValue > 1)                    // if the key is pressed
    {
        tone(speaker, note);           // play a note
        delay(100);                    // wait for 1/10th of a second
    }
    else                               // if the key is not pressed
    {
        noTone(speaker);               // stop playing the note
    }
}
```

Make this change to your code and compile and upload it. Open up the Serial Monitor. Now the printout looks even worse than it did before! Do you have any ideas about why it might look so bad?

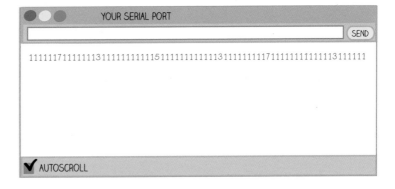

```
void checkPianoKey(int key, int note) {
    touchValue = readCapacitivePin(key); // read touch sensor value
    Serial.print(touchValue);            // send touchValue to the computer
    Serial.print("\t");                  // send a tab
    if (touchValue > 1)                  // if the key is pressed
    {
        tone(speaker, note);             // play a note
        delay(100);                      // wait for 1/10th of a second
    }
    else                                 // if the key is not pressed
    {
        noTone(speaker);                 // stop playing the note
    }
}
```

To fix the problem, add another Serial.print statement telling the LilyPad to print out a tab character after each touchValue. You can do this by adding the line Serial.print("\t"); to your code right after the Serial.print(touchValue); line. Note: The Arduino compiler can't read tabs that you type in directly so the code. "\t" is how you tell Arduino to print a tab.

Compile and upload this code and reopen the Serial Monitor. Things should look a bit better, but you have a display problem that's similar to the last one. Though the numbers have space between them, they're still on one horizontal line and it's hard to figure out which number corresponds to each key. Every time the checkPianoKey procedure is called, it prints out one number followed by a tab, but there is never a carriage return so everything is printed out on one very long line.

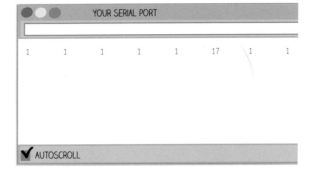

```
void loop() {
  checkPianoKey(key1, 1046);      // play a C
  checkPianoKey(key2, 1175);      // play a D
  checkPianoKey(key3, 1319);      // play a E
  checkPianoKey(key4, 1568);      // play a G
  checkPianoKey(key5, 1760);      // play a A
  checkPianoKey(key6, 2093);      // play a C
  checkPianoKey(key7, 2349);      // play a D
  Serial.println();
  delay(100);
}
```

It would be useful to be able to start the printout on a new line each time the loop section of code runs, because you read in new touchValues for each key each time loop runs, and you want to easily view these updated values. To do this, add one more line of code to the loop section: Serial.println();. This line will send a carriage return to the computer each time loop executes, just after the seven procedure calls. Note that the parentheses after Serial.println(); are empty. This is because you don't want to add any new text to the printout, you just want a new line.

Upload this code to your LilyPad, and open the Serial Monitor. Touch a few different keys while you watch the Monitor. Notice how it's now much easier to determine which number on the printout corresponds to a particular key on your piano.

Save your code.

● ● ●		YOUR SERIAL PORT						
								SEND
17	1	17	1	1	1	1		
17	1	17	1	1	1	1		
17	1	17	1	1	1	1		
17	1	17	1	1	1	1		
17	1	17	1	1	1	1		
17	1	17	1	1	1	1		
17	1	17	1	1	1	1		
17	1	17	1	1	1	1		
17	1	17	1	1	1	1		
✔ AUTOSCROLL								

Printing out your sensor information in this structured way makes troubleshooting much easier. Now, you can see exactly how each sensor is responding. It's easy to tell if a particular sensor isn't working. For instance, if you touch the second key and the second column in your printout doesn't change, you know there's a problem with your second key. You might have an electrical problem, like a loose connection, or a problem with your code—for instance, maybe you're calling checkPianoKey with key1 twice in a row instead of calling it first with key1 and then with key2.

Watching the printout lets you quickly identify the source of many problems. In general, carefully watch your printouts when you're troubleshooting your projects. They can provide very valuable information about the source of errors and well-formatted printouts provide infinitely more useful information than poorly formatted ones do.

CONDENSE YOUR CODE WITH ARRAYS

If you look at the current version of the code, you'll notice a couple of places where similar statements occur several times in a row. For instance, in loop, you call the checkPianoKey procedure seven times in a row, once for each key.

Wouldn't it be nice to be able to condense these lines? To tell the computer that you want to do the same basic thing (call the procedure checkPianoKey) seven times in a row, with seven different sets of variables as input?

```
void loop() {
    checkPianoKey(key1, C);
    checkPianoKey(key2, D);
    checkPianoKey(key3, E);
    checkPianoKey(key4, G);
    checkPianoKey(key5, A);
    checkPianoKey(key6, C1);
    checkPianoKey(key7, D1);
    Serial.println();
    delay(100);
}
```

To accomplish this, it would be useful to have a way to store collections of data. That way you could tell the computer to do checkPianoKey for each element in the collection. One collection could store all of the key information and another could store all of the note information. Essentially, you'd like a way to create a list or table of values, something like this:

KEY LIST →	KEY 1	KEY 2	KEY 3	KEY 4	KEY 5	KEY 6	KEY 7
NOTES LIST →	C	D	E	G	A	C1	D1

Then you could say something like the following: →

```
do the following 7 times:
    checkPianoKey (key list entry1, notes list entry1);
    move to the next entry in both lists
```

ARRAYS: LISTS IN CODE

Arrays allow you to create structured collections of information in code. They're essentially lists or tables in code. Now you're going to create two arrays, one for your keys and one for your notes. To create an array of data, you need to tell the computer a few basic things. First, the **type** of data you're going to be storing in your list, and second, how many elements there are in your list. The format for this statement is shown below.

The statement on the right tells the computer to create an array named keys with places for seven [7] elements and then stores the seven key variables in the array.

TYPE	NAME	#ELEMENTS	=	ELEMENTS IN ARRAY
↓	↓	↓		
int	keys	[7]	=	{key1, key2, key3, key4, key5, key6, key7}

You could make an array for your notes with the statement on the right.

```
int notes[7] = { C, D, E, G, A, C1, D1 };
```

```
int key1 = 6;
int key2 = 9;
int key3 = 10;
int key4 = 11;
int key5 = A2;
int key6 = A3;
int key7 = A4;
int keys[7] = { key1, key2, key3, key4, key5, key6, key7 };

int speaker = 5;
int touchValue;

int C = 1046;
int D = 1175;
int E = 1319;
int F = 1397;
int G = 1568;
int A = 1760;
int B = 1976;
int C1 = 2093;
int D1 = 2349;
int notes[7] = { C, D, E, G, A, C1, D1 };
```

Add array definitions for your keys and notes to the top of your code, in the variable declaration section. Notice that array declarations look kind of like variable declarations. Arrays are variables too. They're just more complex variables than the simple ones you've seen so far. Arrays, like all variables, store information that you can use later on in your program. Compile your code to make sure you haven't made any mistakes in your array definitions.

Before moving on it's worth making a few adjustments. First, notice how you created variables for each key (key1, key2, key3, etc.) and then created the keys array. Well, you only need to do one of these things. You can condense the first section of your code by eliminating the key1, key2, key3, etc. variables and putting the pin numbers for your keys directly into your keys array, as shown below.

```
int key1 = 6;
int key2 = 9;
int key3 = 10;
int key4 = 11;
int key5 = A2;
int key6 = A3;
int key7 = A4;
int keys[7] = { 6, 9, 10, 11, A2, A3, A4 };
```

You can do the same thing for your notes array. Edit your program so that the top of it looks like the code below. Notice how much shorter this is than the code you started with (shown above)!

```
int keys[7] = { 6, 9, 10, 11, A2, A3, A4 };
int notes[7] = { 1046, 1175, 1319, 1598, 1760, 2093, 2349 };
int speaker = 5;
int touchValue;
```

```
const int numberOfKeys = 7;
int keys[numberOfKeys] = { 6, 9, 10, 11, A2, A3, A4 };
int notes[numberOfKeys] = { 1046, 1175, 1319, 1598, 1760, 2093, 2349 };

int speaker = 5;
int touchValue;
```

Finally, notice how you're using the number 7 in both of your array creation statements. This number corresponds to the number of keys on your piano. You can replace this number with an appropriately named variable. Add one more line to the top of your program, creating a new variable called numberOfKeys. Change your array creation statements to make use of it as shown on the left.

Note that you're using `const int` as the data type for the `numberOfKeys` variable instead of `int`. This creates a **constant** variable. This means that the `numberOfKeys` variable will stay constant throughout the program. That is, it will not change. If you try to set `numberOfKeys` to some other value elsewhere in your code, you will get an error when you try to compile and upload your code to the LilyPad.

Now, you'll use your arrays in the rest of your program. Your setup section currently looks like the code shown on the right, with seven `pinMode` statements for the seven piano keys.

You want to replace each `pinMode(key, INPUT);` statement with a statement that uses your new `keys` array. To do so, you need to know how to refer to different elements in the `keys` array. In the first `pinMode` statement you want to use the first element in your `keys` array (instead of `key1`). You want to say:

```
pinMode(1st element of keys array, INPUT);
```

There is a particular code format for accessing elements in an array. You use numbers enclosed in square brackets immediately following the name of the array:

```
keys[element number];
```

With computer code, the numbering is a little...strange. You might think that the first element in an array would be number 1, but it's number 0 instead. Computer scientists start counting with the number 0 instead of the number 1. So, to access the first element in your keys array, you'd say:

```
keys[0]; //access first element in the keys array
```

To access the second element you'd say `keys[1]`, the third, `keys[2]`, and so on. So, the first `pinMode` statement should be: `pinMode(keys[0], INPUT);` Using this new information, you can rewrite your setup section so that it looks like the example shown on the right.

```
void setup() {
    pinMode(speaker, OUTPUT);
    pinMode(key1, INPUT);
    pinMode(key2, INPUT);
    pinMode(key3, INPUT);
    pinMode(key4, INPUT);
    pinMode(key5, INPUT);
    pinMode(key6, INPUT);
    pinMode(key7, INPUT);
    Serial.begin(9600);   // initialize the communication
}
```

```
void setup() {
    pinMode(speaker, OUTPUT);
    pinMode(keys[0], INPUT);
    pinMode(keys[1], INPUT);
    pinMode(keys[2], INPUT);
    pinMode(keys[3], INPUT);
    pinMode(keys[4], INPUT);
    pinMode(keys[5], INPUT);
    pinMode(keys[6], INPUT);
    Serial.begin(9600);   // initialize the communication
}
```

Now turn to the loop section of your program. The first line there looks like this:

```
checkPianoKey(key1, C);
```

Again, instead of using the key variables and notes, you can use your arrays. Here's what a call to checkPianoKey that uses your keys and notes arrays looks like:

```
checkPianoKey(keys[0], notes[0]);
```

You can now replace all of the old calls to checkPianoKey in loop with calls that use your array as shown on the right. Make this final change and try uploading the new code to your LilyPad. Before moving on, make sure that the code compiles and that your piano works nicely. Also open up the Serial Monitor and make sure that your printout still looks good.

```
void loop() {
  checkPianoKey(keys[0], notes[0]);
  checkPianoKey(keys[1], notes[1]);
  checkPianoKey(keys[2], notes[2]);
  checkPianoKey(keys[3], notes[3]);
  checkPianoKey(keys[4], notes[4]);
  checkPianoKey(keys[5], notes[5]);
  checkPianoKey(keys[6], notes[6]);
  Serial.println();
  delay(10);
}
```

CONDENSING THE CODE PART 2: WHILE LOOPS

You may have noticed that, though you've eliminated some variables at the top of your code, you haven't really condensed the loop section. You'll now tackle that challenge.

In your loop section, the calls to checkPianoKey now have a very clear pattern. Each statement is identical except that the numbers you are using to access your arrays increase by one for each line. They start at 0, then advance to 1, then 2, and so on.

You could rewrite the code to capture this pattern by introducing a variable called i and increasing it by one after each checkPianoKey statement. This example is shown on the left. (Note: Don't do this yet, just follow along for a moment.)

```
//create a variable i and initialize it to 0
int i = 0;
checkPianoKey(keys[i], notes[i]);  // i = 0
i = i+1;
checkPianoKey(keys[i], notes[i]);  // i = 1
i = i+1;
checkPianoKey(keys[i], notes[i]);  // i = 2
i = i+1;
checkPianoKey(keys[i], notes[i]);  // i = 3
i = i+1;
checkPianoKey(keys[i], notes[i]);  // i = 4
i = i+1;
...
```

i is initially 0. The first time checkPianoKey is called, the inputs will be keys[0] and notes[0], the second time it's called the inputs will be keys[1] and notes[1] since the variable i is now 1 (i = 0+1). The third time it's called, i will be 2 (i = 0+1+1), so the inputs will be keys[2] and notes[2]. You could repeat this pattern for all seven keys—that is, until i = 6 and the inputs are keys[6] and notes[6].

Notice how, in this example, *every two-line block after the statement* int i = 0; *is identical.*

Rewriting the code this way would increase the number of lines of code in the loop section, which you don't want to do. But, it reveals an intriguing pattern—the fact that, after the statement int i = 0;, two identical lines are repeated seven times.

Now, instead of copying out the same two lines seven times, could you just tell the computer to repeat those lines seven times? You want to say something like the code shown on the right.

```
int i=0;
repeat 7 times:
{
    checkPianoKey(keys[i], notes[i]);
    i = i+1;
}
```

You can use a programming feature, called a while loop, to do this. The term while is used in programming pretty much the same way it is used in real life: "while a certain condition is true, do something." For instance: "While it is raining, stay inside." A while loop in ½ code ½ written English is shown on the left (top).

A real while loop is shown on the left (bottom). This while loop says: while the variable i is less than the numberOfKeys on your piano, do the following:

1. Call the checkPianoKey procedure with keys[i] and notes[i] as inputs.
2. Add 1 to the variable i.

This while loop will repeat until the condition (i < numberOfKeys) is false. Since the variable numberOfKeys is 7, the while loop will repeat seven times. Three snapshots of the while loop execution are shown below, when i=0, when i=3, and when i=7.

```
while (condition)
{
    /* Do what is inside the two curly brackets over and
    over, as long as the condition is true. Stop doing stuff
    only when the condition is not true */

}
```

```
int i = 0;
while (i < numberOfKeys)
{
    checkPianoKey(keys[i], notes[i]);
    i = i+1;
}
```

i=0 CONDITION (0<7) IS TRUE	i=1, i=2, ...	i=3 CONDITION (3<7) IS TRUE	i=4, i=5, i=6 ...	i=7 CONDITION (7<7) IS FALSE

```
while (0 < numberOfKeys)
{
    checkPianoKey(keys[0], notes[0]);
    i = 0+1;
}
```

```
while (3 < numberOfKeys)
{
    checkPianoKey(keys[3], notes[3]);
    i = 3+1;
}
```

```
while (7 < numberOfKeys)
{
    // nothing executes because
    // condition is false
}
```

```
keys[0]  = key1
notes[0] = a C note
```

```
keys[3]  = key4
notes[3] = an E note
```

checkPianoKey is never called

Add this new while loop to the loop section of your code, using it to replace the seven calls to checkPianoKey. Try uploading it to your LilyPad. Your piano should work as it did before, but your program is a little shorter.

More importantly, this new program does a better job of describing what you want your piano to do. It captures information that is lost in the first program, relating the number of times you call checkPianoKey to the number of keys you have in your piano. This code is less error-prone and easier to expand on.

OLD

```
void loop() {
    checkPianoKey(keys[0], notes[0]);
    checkPianoKey(keys[1], notes[1]);
    checkPianoKey(keys[2], notes[2]);
    checkPianoKey(keys[3], notes[3]);
    checkPianoKey(keys[4], notes[4]);
    checkPianoKey(keys[5], notes[5]);
    checkPianoKey(keys[6], notes[6]);
    Serial.println();
    delay(10);
}
```

NEW

```
void loop() {
    int i = 0;
    while (i < numberOfKeys)
    {
        checkPianoKey(keys[i], notes[i]);
        i = i+1;
    }
    Serial.println();
    delay(10);
}
```

A close look at your entire program will reveal another spot where you can replace a large chunk of code with a shorter while loop. Hint: Take a look at the setup section. Can you replace the highlighted portion of code on the right with a while loop? (See the next page for the correct code.)

```
void setup() {
    // set all keys to be inputs
    pinMode(keys[0], INPUT);
    pinMode(keys[1], INPUT);
    pinMode(keys[2], INPUT);
    pinMode(keys[3], INPUT);
    pinMode(keys[4], INPUT);
    pinMode(keys[5], INPUT);
    pinMode(keys[6], INPUT);
    pinMode(speaker, OUTPUT);
    Serial.begin(9600);
}
```

EXPERIMENT

Try associating different notes (frequencies) with different piano keys. Can you get your fabric piano to play a standard scale like the one on a traditional piano? Can you get it to play a short tune when you run your fingers across the keys?

How might you change your code if you wanted to add additional piano keys?

SAVE YOUR CODE

Once you are happy with your program, save it by clicking on the downward pointing arrow in the Toolbar. Your entire program should now look more or less like the one below.

```
const int numberOfKeys = 7;
int keys[numberOfKeys] = { 6, 9, 10, 11, A2, A3, A4 };
int notes[numberOfKeys] = {1046, 1175, 1319, 1598, 1760, 2093, 2349};

int speaker = 5;                          // name of the speaker key
int touchValue;                           // will store sensor readings

void setup() {
    // set all keys to be inputs
    int j = 0;
    while (j < numberOfKeys)
    {
        pinMode(keys[j], INPUT);
        j = j+1;
    }
    pinMode(speaker, OUTPUT);             // set speaker to be an output
    Serial.begin(9600);                   // initialize the communication
}

void loop() {
    int i = 0;
    while (i < numberOfKeys)
    {
        checkPianoKey(keys[i], notes[i]);
        i = i+1;
    }
    Serial.println();
    delay(10);
}

void checkPianoKey(int key, int note) {
    touchValue = readCapacitivePin(key);  // read touch sensor value
    Serial.print(touchValue);             // send value to the computer
    Serial.print("\t");                   // send a tab
    if (touchValue > 1)                   // if the key is pressed
    {
        tone(speaker, note);              // play a note
        delay(100);                       // wait for 1/10th of a second
    }
    else                                  // if the key is not pressed
    {
        noTone(speaker);                  // stop playing the note
    }
}
```

TROUBLESHOOTING, PRINTOUT

CODE PROBLEMS

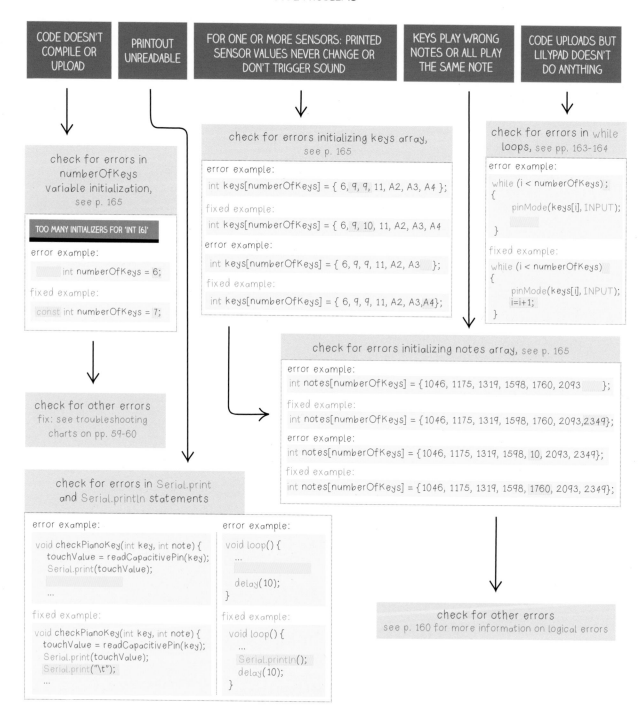

CODE DOESN'T COMPILE OR UPLOAD

check for errors in numberOfKeys Variable initialization, see p. 165

TOO MANY INITIALIZERS FOR 'INT [6]'

error example:

```
         int numberOfKeys = 6;
```

fixed example:

```
const int numberOfKeys = 7;
```

check for other errors
fix: see troubleshooting charts on pp. 59-60

PRINTOUT UNREADABLE

check for errors in Serial.print and Serial.println statements

error example:

```
void checkPianoKey(int key, int note) {
    touchValue = readCapacitivePin(key);
    Serial.print(touchValue);

    ...
```

fixed example:

```
void checkPianoKey(int key, int note) {
    touchValue = readCapacitivePin(key);
    Serial.print(touchValue);
    Serial.print("\t");
    ...
```

error example:

```
void loop() {
    ...


    delay(10);
}
```

fixed example:

```
void loop() {
    ...
    Serial.println();
    delay(10);
}
```

FOR ONE OR MORE SENSORS: PRINTED SENSOR VALUES NEVER CHANGE OR DON'T TRIGGER SOUND

check for errors initializing keys array, see p. 165

error example:

```
int keys[numberOfKeys] = { 6, 9, 9, 11, A2, A3, A4 };
```

fixed example:

```
int keys[numberOfKeys] = { 6, 9, 10, 11, A2, A3, A4
```

error example:

```
int keys[numberOfKeys] = { 6, 9, 9, 11, A2, A3    };
```

fixed example:

```
int keys[numberOfKeys] = { 6, 9, 9, 11, A2, A3,A4};
```

check for errors initializing notes array, see p. 165

error example:

```
int notes[numberOfKeys] = {1046, 1175, 1319, 1598, 1760, 2093        };
```

fixed example:

```
int notes[numberOfKeys] = {1046, 1175, 1319, 1598, 1760, 2093,2349};
```

error example:

```
int notes[numberOfKeys] = {1046, 1175, 1319, 1598, 10, 2093, 2349};
```

fixed example:

```
int notes[numberOfKeys] = {1046, 1175, 1319, 1598, 1760, 2093, 2349};
```

KEYS PLAY WRONG NOTES OR ALL PLAY THE SAME NOTE

CODE UPLOADS BUT LILYPAD DOESN'T DO ANYTHING

check for errors in while loops, see pp. 163-164

error example:

```
while (i < numberOfKeys);
{
    pinMode(keys[i], INPUT);

}
```

fixed example:

```
while (i < numberOfKeys)
{
    pinMode(keys[i], INPUT);
    i=i+1;
}
```

check for other errors
see p. 160 for more information on logical errors

MAKE BEAUTIFUL MUSIC

The LilyPad speaker is fun, but frankly it's pretty limited. It's not very loud, it can only play one note at a time, and it can only produce simple beeping sounds. What if you wanted to play loud beautiful sounds using high-quality speakers? What if you wanted to play chords? What if you wanted your piano to actually sound like a piano? Well, one good way to expand the musical range of your piano is to use it as a controller for a computer and use the computer to generate sounds. That's what you'll do in this final section.

PLAYING MUSIC ON YOUR COMPUTER

Before you begin this section, make sure you've gone through all of the previous parts of this chapter. The computer-controlled piano will only work if your printouts are properly formatted.

Download the Piano Application for your Mac or PC here: http://www.sewelectric.org/PianoApplication. Follow the installation instructions on the web page to get the application installed and running. Note: The Piano Application may not run on all computers. See http://www.sewelectric.org/PianoApplication for more information.

Once you have installed the application, a window asking you to choose a serial port should pop up. Choose the correct serial port from the list. This should be the same serial port you selected in the Arduino software.

Try playing your piano now. You should hear piano tones coming from your computer and see waves on the application window like the image shown on the right (bottom). These waves are showing you the frequency of the notes you're playing on your piano.

Note: If you are unable to download, install, or run the application, see http://www.sewelectric.org/PianoApplication for troubleshooting information.

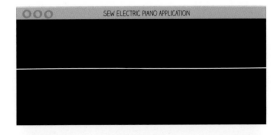

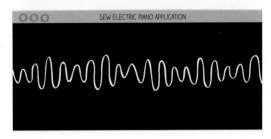

The application that's running on the computer is doing pretty much the same thing that the checkPianoKey procedure you wrote was doing. It is using the information about whether or not a particular key is pressed to determine whether it should make a particular sound. More specifically, it is looking at the stream of sensor data you are sending to the computer with your Serial.print and Serial.println statements and using it to detect when a key is pressed. If a key is pressed, the application generates a piano note. Like the program you wrote, it generates a different note for each key. If you are interested in seeing the code for this program, you can download it here: http://www.sewelectric.org/PianoApplicationCode.

You've probably noticed that there's something annoying about the current situation: Both your fabric piano and the computer are making sounds. You'll want to create a new version of your Lily-Pad Arduino program where the LilyPad speaker stays quiet so that you can clearly hear the sounds being made by the computer.

Quit the Piano Application. Open up your Piano program in the Arduino software. Go to the File menu and select "Save As...". Choose a new name for the code that will work with the Piano Application, something like "pianoForComputer".

Look at your checkPianoKey procedure to find what you need to change to get rid of the sounds being made by the LilyPad speaker. The lines that check whether keys are being pressed and generate sounds are highlighted on the right (top).

Since this work is now being done by the computer, you can delete this entire section of the checkPianoKey procedure, making it much simpler, like the version shown on the right (bottom).

All the procedure needs to do now is to read sensor information from the piano key and send the reading to the computer. The computer does everything else.

```
void checkPianoKey(int key, int note) {
    touchValue = readCapacitivePin(key);   // read touch sensor value
    Serial.print(touchValue);               // send value to the computer
    Serial.print("\t");                     // send a tab
    if (touchValue > 1)                     // if the key is pressed
    {
        tone(speaker, note);                // play a note
        delay(100);                         // wait for 1/10th of a second
    }
    else                                    // if the key is not pressed
    {
        noTone(speaker);                    // stop playing the note
    }
}
```

```
void checkPianoKey(int key, int note) {
    touchValue = readCapacitivePin(key);   // read touch sensor value
    Serial.print(touchValue);               // send value to the computer
    Serial.print("\t");                     // send a tab
}
```

Upload this new code to your computer, restart the Piano Application, and play with your piano. Note: If you opened the Serial Monitor in Arduino you should close it before you open the Piano Application. Your piano should sound nicer now. In Arduino, save the changes you've made to this "pianoForApplication" version of the code.

If you want to unplug your piano from the computer and have it make sounds with the speaker, upload your original "Piano" program to the LilyPad. You can find the original program in your Arduino Sketchbook. Click on the upward pointing arrow, and choose "Piano" (or the name that you gave your original program). Then upload this program to your LilyPad.

Note: If you have trouble downloading, installing, or using the Piano application, see http://www.sewelectric.org/PianoApplication for troubleshooting information.

PLAY!

Now you've got a nice sounding instrument that you made! Compose a short song and play it for your friends and family. Let them play your piano. Start a band!

WASH

If your piano gets dirty, you can hand wash it in cold water. Remember to take the LilyPad SimpleSnap off first; otherwise you can damage the battery and the LilyPad.

RECHARGE

If your piano stops playing when it's not attached to the computer, its battery has probably died. Attach your LilyPad to your computer to recharge it.

TROUBLESHOOTING SOLUTIONS
ELECTRICAL PROBLEMS

Electrical problems are problems with either the design or construction of your electrical circuits. These are physical problems that you'll need to fix by hand with scissors, a needle, thread, fabric, and glue. The next several pages provide examples of common electrical problems, information about how to find them, and step-by-step instructions for fixing them.

ELECTRICAL PROBLEMS COVERED IN THIS SECTION

• loose connections
• short circuits
• reversed polarity

LOOSE CONNECTIONS

Loose connections occur when the thread that is stitched through a component (like an LED, speaker, LilyTiny, or Protoboard) is too loose. If the thread is too loose, there will not be a consistent electrical connection between the thread and the component. To carry electricity through the circuit, the thread must be tightly pressed up against the silver holes in the component. Loose connections can be caused by loose stitching or unravelling knots. If a knot is not secured with glue, it can come undone, loosening the connections near it.

symptoms
If there is a loose connection in your project, parts of your project will only work some of the time. For example, if there is a loose connection between your LED and your battery board in the bookmark project, your LED may flicker on and off or only work some of the time.

checking your project for loose connections
Gently bend and stretch your project. If this causes your LED to turn on and off, it is likely that you have a loose connection. To find loose connections, look carefully at the connections between your thread and your components. Make sure each connection is snug and tight. If you find a loose thread around any tab, or an unravelling knot, you'll need to fix it.

FIX LOOSE CONNECTIONS

To fix loose connections, thread your needle with conductive thread. From the back or underside of the fabric, push your needle up through the tab with the loose connection. Loop through the tab a few times. Make sure your thread is touching the original stitches you sewed in several places. Push the needle to the back or underside of the fabric. Tie a snug knot, making sure that the new thread is pulled tightly against the tab and the old stitching.

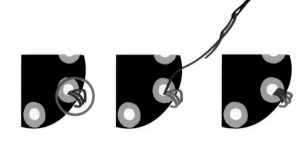

If you have a knot that is unravelling, find the
end of the thread and pull on it to re-tighten
connections. Cut out a small piece of fabric
and glue it down over the unravelling thread.
You will also need to resew stitches that
have come undone. Make sure that your
new thread touches the existing thread in
several places to make a solid electrical
connection.

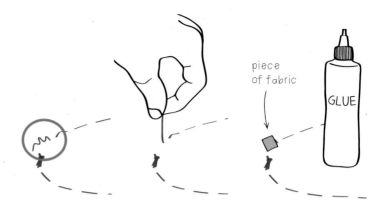

piece
of fabric

GLUE

SHORT CIRCUITS

Short circuits or "shorts" happen when two threads that should not touch one another come into contact. Power (+) and ground
(-) traces in a circuit should never touch one another. More generally, traces connected to different tabs on a Protoboard, LilyTiny,
or other component should never touch one another. It is OK for traces connected to the same tab to touch each other. For
instance, all of the threads attached to the ground (-) tab can touch each other, but thread attached to tab 9 should never touch
thread attached to the ground (-) tab.

Causes of shorts include: long knot tails that brush up against each other, stitches from two different tabs that cross each other,
stitching that continues from the (+) tab of a component to the (-) tab of the same component (That is, you forgot to stop and tie a
knot after stitching through one of its tabs.), and loose thread around a tab that brushes against a neighboring tab. Each of these
problems is covered in more detail later in this section.

symptoms
If there is a short circuit in your project, parts of your project either will not work or will only work some of the time. For example,
if you have sewn across your LED in the bookmark or monster project, forgetting to stop your stitching after the (+) or (-) tab,
your LED will not work. If you left long knot tails dangling behind your speaker in the monster project, the speaker will stop working whenever the tails brush up against one another. This will result in a speaker that works some of the time but not others.

checking your project for short circuits
To find shorts, follow each trace in your project looking for the problems described in this section. Check all of your knot tails to
make sure that they can't touch other traces, make sure that none of your traces cross or touch other traces, make sure that you
have not sewn across any of your components, and, make sure that there are no loose threads that may touch other traces or
tabs.

FIX SHORT CIRCUITS

Loose thread touching another pin or trace:
From the back of your fabric (if possible), pull on
the loose thread until it is gathered in one spot. Cut
out a small piece of fabric and glue it down over the
extra thread, making sure that the extra thread is
not touching any neighboring traces or tabs.

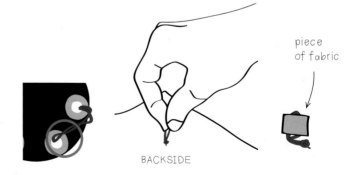

piece
of fabric

BACKSIDE

Stitching across a component: If you have accidentally sewn the (+) side of a component to the (-) side of a component, you will need to cut the thread that is connecting the two tabs. You will be left with two very short threads. Tug on each thread tightly, move it away from the other thread, and glue it down with a small piece of fabric.

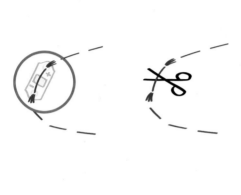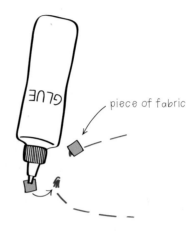

piece of fabric

Knot tails touching: To eliminate long knot tails, trim them down to 1/4" (6mm) or shorter. Seal them with glue so that they do not come unravelled. Make sure that the tails cannot brush up against any neighboring traces. You can also glue a small piece of fabric down over your knot tails to keep them from unravelling and touching other traces.

Overlapping stitches: If there is an area in your project where two traces cross each other and come into contact, separate the two traces with a small piece of fabric. Slide the piece of fabric underneath one set of stitches and on top of the other to keep them apart. Glue the fabric in between the two threads so that they cannot touch one another.

piece of fabric

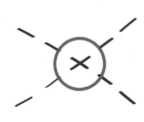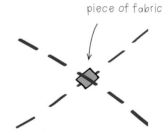

REVERSED POLARITY

When you stitch the (+) side of a component to the (-) side of a battery (or the (-) tab of your Protoboard), and the (-) side of a component to the (+) side of a battery (or one of the numbered tabs on your Protoboard), you have connected it backwards and it will not work. Electricity will only flow through most components in one direction—from the (+) to the (-) side of a component. Examples of reversed polarity are described below.

symptoms
If one of your components is sewn in backwards, it will not work at all. For instance, if you've sewn your LED on backwards in the bookmark project it will not turn on.

checking your project for reversed polarity
To find instances of reversed polarity, look carefully at each component in your project. Make sure that each component's (-) tab is stitched to a matching (-) tab on the battery holder, LilyTiny, or Protoboard. Similarly, make sure that each component's (+) tab is sewn where it should be—to the (+) tab on a battery holder or one of the numbered tabs on the LilyTiny or Protoboard.

FIX REVERSED POLARITY

Unfortunately there's no easy way to fix instances of reversed polarity. To correct this problem you need to remove the component and reattach it in the correct orientation. You'll need to cut your stitches to remove the component. Once you've removed the component, glue it back on in the correct orientation. Make sure you get it right this time!

Thread your needle with conductive thread. Begin with the (-) trace. On the back or underside of the fabric (if possible), tie your new thread to the old stitching on the (-) trace. Sew toward the reattached component. Make sure your new thread is touching the original stitches in several places. Stitch through the (-) tab of the component several times. Tie a knot on the underside of your fabric, trim its tails, and seal it with a dab of glue. Repeat the same process for the (+) tab of your component.

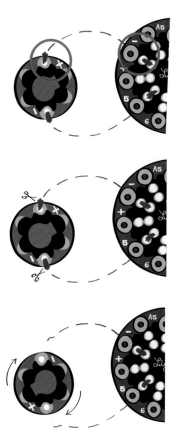

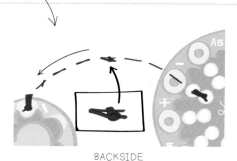

BACKSIDE

TROUBLESHOOTING SOLUTIONS
CODE PROBLEMS, COMPILE ERRORS

Compile errors are errors that happen when you attempt to compile your code and the Arduino software (the compiler) finds an error in it. If you receive an error when you attempt to compile and upload your code, you know you have either a compile error or an upload error.

symptoms
If you're not sure if you have a compile error or an upload error, scroll to the top of the black area at the bottom of the Arduino window. If you see the name of your program followed by .ino in orange text at the top of the box, you know you have a compile error (right, top). If you see a message in white text at the top of the box that looks something like "Binary sketch size: 4,962 bytes (of a 30,720 byte maximum)" you know that you have an upload error (right, bottom).

Compile error:

> A FUNCTION-DEFINITION IS NOT ALLOWED HERE BEFORE '{' TOKEN
>
> Blink.ino: In function 'void setup()':
> Blink:19: a function-definition is not allowed here before '{' token
> Blink:24: error: expected '}' at end of input

Upload error:

> SERIAL PORT '/DEV/TTY.BLUETOOTH-MODEM' ALREADY IN USE. TRY QUITTING ANY PROGRAMS THAT MAY ...
>
> Binary sketch size: 4,962 bytes (of a 30,720 byte maximum)
> processing.app.SerialException: Serial port '/dev/tty.Bluetooth-Modem' already in use....
>   at processing.app.Serial.<init>(Serial.java:171)
>   at processing.app.Serial.<init>(Serial.java:77)

COMPILE ERRORS COVERED IN THIS SECTION

- missing semicolons

- missing curly brackets

- missing parentheses

- missing commas

- misspellings and mis-capitalizations

- missing variable definitions

- random extra text in program

finding compile errors in your code
Compile errors often occur because of missing or incorrect punctuation, misspellings, and mis-capitalizations. They will also occur if you attempt to use a variable you have not initialized at the top of your program or if there is extra text anywhere in your program. If you get a compile error, begin by carefully reading the error messages in the feedback area of the Arduino window for clues. Investigate your code around the location that the Arduino software highlights or jumps to. If you cannot see any problems in that area, carefully read through your entire program line by line, looking for the problems described in this section.

the rest of this section
The next few pages explore common code mistakes (the ones listed on the left) and the compile errors they generate, along with tips on how to find and fix these problems.

Each section begins with examples of problematic pieces of code. The location of the problem is highlighted in yellow. Below each code example is an example of the kind of error you might receive if you made a similar mistake in your code. This is followed by advice on finding and fixing similar errors.

MISSING SEMICOLONS ;

```
int led = 13

void setup() {
    pinMode(led, OUTPUT);
}
```

EXPECTED UNQUALIFIED-ID BEFORE NUMERIC CONSTANT

```
Blink:11: error: expected unqualified-id before numeric constant
Blink:13: error: expected ',' or ';' before 'void'
```

```
void loop()  {
    digitalWrite(led, HIGH);
    delay(1000);
    digitalWrite(led, LOW); .
    delay(1000)
}
```

EXPECTED ',' OR ';' BEFORE '}' TOKEN

```
Blink.ino: In function 'void loop()':
Blink:24: error: expected ';' before '}' token
```

```
if (touchValue > 1)
{
    tone(speaker, C)
    delay(100);
}
```

EXPECTED ',' OR ';' BEFORE 'DELAY'

```
piano.ino: In function 'void checkPianoKey(int, int)':
piano:37: error: expected ';' before 'delay'
```

symptoms
Error messages for missing semicolons are relatively straightforward. In each error message, either in the orange or the black feedback area, there is a line that says: error: expected ';' before... This is an indication that you're missing a semicolon. The Arduino software will usually highlight the line immediately after the missing semicolon to indicate the location of your error.

FIX MISSING SEMICOLONS ;

Look for instances of error: expected ';' in your error. Look for similarities between your error and the ones above. Replace the missing semicolon and recompile your code. If the missing semicolon was the only compile error in your program, your code will now compile. If you have additional errors, you will get a new error message with information about the next error in your program.

MISSING CURLY BRACKETS { }

```
void setup() {
    pinMode(led, OUTPUT);
```

A FUNCTION-DEFINITION IS NOT ALLOWED HERE BEFORE '{' TOKEN

```
Blink.ino: In function 'void setup()':
Blink:19: a function-definition is not allowed here before '{' token
Blink:24: error: expected '}' at end of input
```

```
void setup()
    pinMode(led, OUTPUT);
}
```

EXPECTED INITIALIZER BEFORE 'PINMODE'

```
Blink:16: error: expected initializer before 'pinMode'
Blink:17: error: expected declaration before '}' token
```

```
if (touchValue > 1)

    tone(speaker, C);
    delay(100);
}
```

EXPECTED UNQUALIFIED-ID BEFORE 'ELSE'

```
piano:18: error: expected unqualified-id before 'else'
piano:19: error: expected unqualified-id before 'else'
piano:20: error: expected unqualified-id before 'else'
piano:21: error: expected unqualified-id before 'else'
piano:39: error: expected unqualified-id before 'else'
piano:43: error: expected declaration before '}' token
```

symptoms
These error messages are often confusing and cryptic. Most will say something about either a '{' or '}' somewhere in the orange or black feedback area. The Arduino software will sometimes highlight the line immediately after the missing curly bracket to indicate the location of your error. Sometimes, unfortunately, it will highlight a completely unrelated line.

FIX MISSING CURLY BRACKETS { }

Look for mentions of '{' or '}' in your errors. Look for similarities between your error and the ones above. Replace the missing bracket and recompile your code. If you have no additional compile errors in your code, your code will compile successfully, otherwise you'll get a new error message with information about the next error in your program.

MISSING PARENTHESES ( )

```
void setup( {
  pinMode(led, OUTPUT);
}
```

VARIABLE OR FIELD 'SETUP' DECLARED VOID

Blink:14: error: variable or field 'setup' declared void
Blink:14: error: expected primary-expression before '{' token

```
void loop()  {
  digitalWrite(led, HIGH);
  delay(1000  ;
  digitalWrite(led, LOW);
  delay(1000);
```

EXPECTED ')' BEFORE ';' TOKEN

Blink.ino: In function 'void loop()':
Blink:22: error: expected ')' before ';' token

```
void checkPianoKey (int key, int note  {
  touchValue = readCapacitivePin(key);
  Serial.print(touchValue);
  Serial.print("\t");
  ...
```

'CHECKPIANOKEY' WAS NOT DECLARED IN THIS SCOPE

piano.ino: In function 'void loop()':
piano:23: error: 'checkPianoKey' was not declared in this scope
piano.ino: At global scope:
piano:30: error: expected ')' before '{' token

symptoms
These error messages are often confusing and cryptic. Some, but not all, will mention a '(' or ')' token in either the orange or black feedback area. The Arduino software will sometimes highlight the line with the missing parenthesis to indicate the location of your error, but sometimes it will highlight an unrelated line.

FIX MISSING PARENTHESES ( )

Look for mentions of '(' or ')' in your errors. Look for similarities between your error and the ones above. Replace the missing parenthesis and recompile your code. If you have no additional compile errors in your code, your code will compile successfully, otherwise you'll get a new error message with information about the next error in your program.

MISSING COMMAS ,

```
void setup() {
  pinMode(led  OUTPUT);
}
```

EXPECTED ')' BEFORE NUMERIC CONSTANT

Blink.ino: In function 'void setup()':
Blink:16: error: expected ')' before numeric constant
/Applications/Arduino.app/Contents/Resources/Java/...
Blink:16: error: at this point in file

```
while (i < numberOfKeys)
{
  checkPianoKey(keys[i]  notes[i]);
  i = i+1;
}
```

EXPECTED ')' BEFORE 'NOTES'

piano.ino: In function 'void loop()':
piano:23: error: expected ')' before 'notes'
piano:4: error: too few arguments to function
'void checkPianoKey(int, int)'
piano:23: error: at this point in file

```
const int numberOfKeys = 7;
int keys[numberOfKeys] = { 6, 9  10, 11, A2, A3, A4 };
```

EXPECTED ')' BEFORE NUMERIC CONSTANT

piano:2: error: expected '}' before numeric constant
piano:2: error: expected ',' or ';' before numeric constant
piano:2: error: expected declaration before '}' token

symptoms
These error messages are often confusing and cryptic. Very few will mention a comma in the orange or black feedback area. The Arduino software will usually highlight the line with the missing comma to indicate the location of your error.

FIX MISSING COMMAS ,

Look for similarities between your error and the ones above. Replace the missing comma and recompile your code. If you have no additional compile errors in your code, your code will compile successfully, otherwise you'll get a new error message with information about the next error in your program.

MISSPELLINGS AND MIS-CAPITALIZATIONS

```
void loop()  {
  digitalWrite(led, HIGH);
  delay(1000);
  digitalWrite(led, LOW);
  dellay(1000);
}
```

'DELLAY' WAS NOT DECLARED IN THIS SCOPE

```
Blink.cpp: In function 'void loop()':
Blink:21: error: 'dellay' was not declared in this scope
```

```
void loop()  {
  digitalWrite(led, HIGH);
  delay(1000);
  digitalWrite(led, low);
  delay(1000);
}
```

'LOW' WAS NOT DECLARED IN THIS SCOPE

```
Blink.ino: In function 'void loop()':
Blink:23: error: 'low' was not declared in this scope
```

```
if (touchValue < 1000)
{
    Song(2000);
}
```

'SONG' WAS NOT DECLARED IN THIS SCOPE

```
monster.ino: In function 'void loop()':
monster:29: error: 'Song' was not declared in this scope
```

symptoms

Error messages for misspellings and mis-capitalizations are some of the clearest and most helpful you'll encounter. All take the form of 'word with error' was not declared in this scope. The Arduino software will almost always highlight the line with the misspelling or mis-capitalization to indicate the location of your error. Also, if you have misspelled one of the built-in Arduino procedures or variables, the color of the misspelled or mis-capitalized word will change from orange or blue to black, giving you another valuable clue about the cause of the error.

FIX MISSPELLINGS AND MIS-CAPITALIZATIONS

Look for errors of the form 'word with error' was not declared in this scope. Correct the misspelling or mis-capitalization and recompile your code. If you have no additional compile errors in your code, your code will compile successfully, otherwise you'll get a new error message with information about the next error in your program.

MISSING VARIABLE DEFINITIONS

```
// missing int led=13; line

void setup()
  pinMode(led, OUTPUT);
}
```

'LED' WAS NOT DECLARED IN THIS SCOPE

```
Blink.ino: In function 'void setup()':
Blink:4: error: 'led' was not declared in this scope
Blink.ino: In function 'void loop()':
Blink:9: error: 'led' was not declared in this scope
```

```
// missing const int numberOfKeys = 7; line

int keys[numberOfKeys] = { 6, 9, 10, 11, A2, A3, A4 };
```

'NUMBEROFKEYS' WAS NOT DECLARED IN THIS SCOPE

```
piano:2: error: 'numberOfKeys' was not declared in this scope
piano:3: error: 'numberOfKeys' was not declared in this scope
piano.ino: In function 'void setup()':
piano:11: error: 'numberOfKeys' was not declared in this scope
...
```

```
int led = A4;
int speaker = 5;
// missing int aluminumFoil=A2; line
int sensorValue;
```

'ALUMINUMFOIL' WAS NOT DECLARED IN THIS SCOPE

```
monster.ino: In function 'void setup()':
monster:18: error: 'aluminumFoil' was not declared in this scope
monster.ino: In function 'void loop()':
monster:24: error: 'aluminumFoil' was not declared in this scope
```

symptoms

Errors for missing variable definitions are fairly clear. All take the form of 'missing variable' was not declared in this scope. The Arduino software will highlight the first line in your program that uses the missing variable. For example, in the code above on the far left, Arduino will highlight the pinMode(led, OUTPUT); line when you attempt to compile the code since it is the first line that uses the missing variable led.

FIX MISSING VARIABLE DEFINITIONS

If you receive errors in the form of 'missing variable' was not declared in this scope, check for missing or misspelled variable definitions at the top of your code. Correct or add the necessary definitions. Once you fix the problem, if you have no additional compile errors in your code, your code will compile successfully. Otherwise, you'll get a new error message with information about the next error in your program.

RANDOM EXTRA TEXT IN PROGRAM

```
void loop()   {
  digitalWrite(led, HIGH);   x
  delay(1000);
  digitalWrite(led, LOW);
  delay(1000);
}
```

'X' WAS NOT DECLARED IN THIS SCOPE

```
Blink.ino: In function 'void loop()':
Blink:9: error: 'x' was not declared in this scope
Blink:10: error: expected ';' before 'delay'
```

```
void setup() {
  / set all keys to be inputs
  int i = 0;
  while (i < numberOfKeys)
  {
    ...
```

EXPECTED PRIMARY-EXPRESSION BEFORE '/' TOKEN

```
piano.ino: In function 'void setup()':
piano:9: error: expected primary-expression before '/' token
piano:9: error: 'set' was not declared in this scope
piano:9: error: expected ';' before 'all'
piano:11: error: 'i' was not declared in this scope
```

```
void song(int duration) {
  tone(speaker, C);
  delay(duration);
  tone(speaker, D)k;
  delay(duration);
}
```

EXPECTED ';' BEFORE 'k'

```
monster.ino: In function 'void song(int)':
monster:40: error: expected ';' before 'k'
```

symptoms

Errors for extra text in your program can be confusing, but Arduino will usually highlight the line where the extra text is, making these problems easier to find. A common extra text error is a mistyped comment—a comment that is missing a '/', '/*' or '*/'. An example of this kind of error is shown above in the middle. Here the comment is missing one of its beginning slash '/' characters. If you have a mistyped comment, its color will be black instead of grey, giving you a clue to the problem.

FIX RANDOM EXTRA TEXT IN PROGRAM

Correct your mistyped comment or remove the extra text from your code. When you recompile, if you have no additional compile errors, your code will compile successfully. If you have additional issues you'll get a new error message with information about the next thing you need to fix.

TROUBLESHOOTING SOLUTIONS
CODE PROBLEMS, LOGICAL ERRORS

Logical errors occur when your code compiles and uploads, but doesn't behave the way you want it to. These errors are the trickiest to find and fix because the Arduino software doesn't give you any feedback about what might be causing the problem, like it does with compile and upload errors.

Logical errors happen when the code that you've written doesn't do what you intended it to do. The code compiles and the LilyPad is following the code exactly as you've written it, but there is an error in the code that is making your project behave in an unexpected way.

symptoms

If your project isn't working the way you expect it to, you may have a logical error in your code. For instance, if your code compiles and uploads but your project doesn't do anything, or if (in the monster project) your monster doesn't ever stop making sound, you might have a logical error. You might, however, also have an electrical problem.

finding logical errors in your code

Before you check your code for logical errors, you should make sure your project doesn't have any electrical problems. Once you've eliminated the possibility of electrical issues, you should check your code for errors.

To find logical errors, read through your program line by line and try to relate what each line is doing to the behavior you're seeing in your project. The LilyPad does exactly what your program tells it to do. Your job is to find the mismatch between what you want it to do and what the program is telling it to do.

In your read-through, look especially for the errors listed on the left. However, keep in mind that there is an infinite variety of logical errors and this guide cannot cover them all. Your particular error may not be discussed here.

the rest of this section

The next few pages explore common logical errors—the ones listed on the left—and the behaviors they generate. This section also provides tips on how to find and fix these problems.

Each section begins with examples of code with and without the specific error. The location of the problems and corrections are highlighted in yellow in the code. Below the code examples is a discussion of the behavior you will see if you make a similar mistake in your code. This is followed by advice on finding and fixing that particular type of problem.

LOGICAL ERRORS COVERED IN THIS SECTION

• missing variable initializations

• variables that do not match circuit

• conditions that are always true or false

• extra or misplaced semicolons

• incorrect variable initializations

• problems with delay statements

MISSING VARIABLE INITIALIZATIONS

error example:

```
int led = 13;

void setup() {

}
```

fixed example:

```
int led = 13;

void setup() {
  pinMode(led, OUTPUT);
}
```

error example:

```
void setup() {
  pinMode(led, OUTPUT);
  pinMode(speaker, OUTPUT);
  pinMode(aluminumFoil, INPUT);

  Serial.begin(9600);
}
```

fixed example:

```
void setup() {
  pinMode(led, OUTPUT);
  pinMode(speaker, OUTPUT);
  pinMode(aluminumFoil, INPUT);
  digitalWrite(aluminumFoil, HIGH);
  Serial.begin(9600);
}
```

symptoms

If you have an LED that is turning on very dimly or a speaker that is making very faint sounds, you may be missing a pinMode(component, OUTPUT); statement like the example above on the top left. If you are missing a pinMode statement for a component, the component will either not work at all, or will barely work. For example, if the component is an LED it will only turn on very dimly.

If you have a sensor or switch that is giving you random readings, you may be missing a digitalWrite(component, HIGH); statement like the example above on the bottom left. If you're missing this digitalWrite(component, HIGH); statement, your sensor or switch will not work properly. It will behave erratically and you will not be able to get reliable readings from it.

FIX MISSING VARIABLE INITIALIZATIONS

If you are experiencing anything like the problems described above, look for missing variable initialization statements in setup. Remember that each component you use in your project needs to be initialized in setup. You should have a pinMode statement for each LED, speaker, sensor, or switch in your design. Each sensor or switch (with the exception of the capacitive sensors in the piano project) should also have a digitalWrite(component, HIGH); statement in setup. If you find missing initialization statements, add them to your program and recompile and upload your code. Test your project to see if you have fixed the error.

VARIABLES THAT DO NOT MATCH CIRCUIT

error example:
`int speaker = 5;`

fixed example:
`int speaker = 6;`

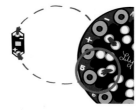

error example:
`int led = A4;`

fixed example:
`int led = 5;`

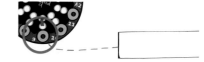

error example:
`int key1 = 6;`

fixed example:
`int key1 = 9;`

symptoms

If there is an LED, speaker, or sensor in your project that is not working at all, your code may not match the circuit that you have built. For example, if your LED is sewn to pin 5, but your code is written as though your LED is sewn to pin A4, your LED will not work. This is the example shown above in the center.

FIX VARIABLES THAT DO NOT MATCH CIRCUIT

If you have a component that is not working at all, look carefully at your circuit and see which pins each of your components are sewn to. Make sure your code matches your circuit. If you find any places where circuit and code disagree, you've identified your problem.

You can address this problem in two different ways. You can either restitch your circuit so that it matches the code you've written or you can rewrite your code so that it matches the circuit that you've sewn. It's a lot easier to change your code than to restitch your circuit! If you can, you want to edit your code so that it matches your circuit. For example, if the (+) side of your speaker is sewn to pin 6, change your code so that the `speaker` variable in your code also refers to pin 6 (as in the left-hand example shown above).

There is one exception to this general rule. For the monster project's sensor, one paw should be sewn to (-) and the other paw should be sewn to one of the "A" pins—pin A2, A3, A4, or A5. If you sewed your sensor paw to a pin that is not A2, A3, A4, or A5, you will need to take out your stitching and resew the sensor paw to one of the "A" pins. In all other cases, you should change your code instead of changing your circuit to correct the mismatch.

Note: See also "incorrect variable initializations" on page 165.

CONDITIONS THAT ARE ALWAYS TRUE OR FALSE

error example:

```
sensorValue = analogRead(aluminumFoil);
Serial.println(sensorValue);
delay(100);
if (sensorValue < 1024)
{
    song(2000);
}
```

fixed example:

```
sensorValue = analogRead(aluminumFoil);
Serial.println(sensorValue);
delay(100);
if (sensorValue < 1000)
{
    song(2000);
}
```

error example:

```
const int numberOfKeys = 7;
int i = 0;
while (i < numberOfKeys)
{
    pinMode(keys[i], INPUT);

}
```

fixed example:

```
const int numberOfKeys = 7;
int i = 0;
while (i < numberOfKeys)
{
    pinMode(keys[i], INPUT);
    i = i+1;
}
```

symptoms

If something that should happen in your program never happens—for example, if your monster never sings its song—or, if something happens all the time when it shouldn't—for example, if your monster never stops singing—you may have a bad conditional statement. You may have a conditional statement with a condition that is always true or always false. For example, in the code above on the top left, the condition (sensorValue < 1024) will always be true because the variable sensorValue will always be below the threshold value of 1024. This means your monster will always be singing; it will never stop singing.

If your program uploads but your project doesn't do anything at all, you may also have a bad conditional statement. In the bottom left example above, (i < numberOfKeys) is always true because i never changes, i is always 0. The while statement is missing an i=i+1; statement. This means you will be stuck inside the while loop forever; the LilyPad will never move on to the rest of your program.

FIX CONDITIONS THAT ARE ALWAYS TRUE OR FALSE

If you are experiencing anything like the problems described here, look carefully at the conditional statements in your program—if else statements and while loops. Read through these sections of your program and make sure that the conditions in each conditional statement will change as the program progresses or as you interact with your project.

Once you've identified the problematic condition, adjust it so that it will change as the program progresses or you interact with your project. Make sure that your threshold values are appropriate for the sensor values in your project and make sure that you are not getting stuck in while loops because of missing i=i+1; statements.

Note: See also the extra semicolons entry on page 164. Extra and/or misplaced semicolons can create similar behaviors and problems.

EXTRA OR MISPLACED SEMICOLONS

error example:

```
if (touchValue > 1);
{
    tone(speaker, note);
    delay(100);
}
```

fixed example:

```
if (touchValue > 1)
{
    tone(speaker, note);
    delay(100);
}
```

error example:

```
const int numberOfKeys = 7;
int i = 0;
while (i < numberOfKeys);
{
    checkPianoKey(keys[i], notes[i]);
    i = i+1;
}
```

fixed example:

```
const int numberOfKeys = 7;
int i = 0;
while (i < numberOfKeys)
{
    checkPianoKey(keys[i], notes[i]);
    i = i+1;
}
```

symptoms

If something in your program happens all the time when it shouldn't—for example, if your monster never stops singing—you may have an extra semicolon in one of your if statements. If your program uploads but your project doesn't do anything at all, you may have an extra semicolon in one of your while statements.

When you put a semicolon at the end of the first line of a conditional statement, as in the left-hand examples above, your code will not behave the way you expect it to. Adding this extra semicolon tells the compiler that the entire conditional statement should end where the semicolon is. Anything after the semicolon is treated as a separate and unrelated block of code.

For example, in the code on the top left above, the entire if statement would end with the semicolon. The statements tone(speaker, note); and delay(100); would be treated as independent statements, unrelated to the if statement. They would always be executed, whether or not the condition in the if statement was true. This means the speaker would always play a note. Note: This particular problem will only happen with stand-alone if statements. If your if statement has a companion else statement, you will receive a compile error when you try to compile the code with the extra semicolon.

In the code on the bottom left above, the entire while loop would end with the semicolon. Because there is no statement that changes the value of i between the while loop's condition and the semicolon, the condition of the while loop would never change and the LilyPad would remain stuck in this part of the code forever. The code would upload to your project, but your project wouldn't do anything once it reached the while loop with the extra semicolon.

FIX EXTRA OR MISPLACED SEMICOLONS

If you are experiencing anything like the problems described here, look carefully at the conditional statements in your program—if statements and while loops. Read through these sections of your program and make sure that they do not have any extra semicolons. The lines if (condition) and while (condition) should never be followed by semicolons. Remove any extra semicolons you find and recompile and upload your code.

INCORRECT VARIABLE INITIALIZATIONS

error example:

```
int C = 10;
int D = 1175;
int E = 1319;
int F = 1397;
int G = 15;
int A = 17600;
int B = 1976;
int C1 = 2093;
```

fixed example:

```
int C = 1046;
int D = 1175;
int E = 1319;
int F = 1397;
int G = 1568;
int A = 1760;
int B = 1976;
int C1 = 2093;
```

error example:

```
const int numberOfKeys = 7;
int i = 1;
while (i < numberOfKeys)
{
    checkPianoKey(keys[i], notes[i]);
    i = i+1;
}
```

fixed example:

```
const int numberOfKeys = 7;
int i = 0;
while (i < numberOfKeys)
{
    checkPianoKey(keys[i], notes[i]);
    i = i+1;
}
```

symptoms

Variables, their definitions and initializations, lay the foundation for the rest of your program. They're like the ingredients in a recipe, the basic raw materials you have to work with. If you give a variable the wrong value at the beginning of your program it's like starting out with the wrong ingredient for your recipe (salt instead of sugar, say). Variables with bad values can cause lots of different kinds of problems in your programs.

A pin variable that does not match your circuit (an error covered earlier in this section, on page 162) is one specific example of an incorrect variable initialization. In this case, the bad variable initialization means that one of your components will not work properly.

The example above on the left shows a different example, the incorrect initialization of several note variables. This will cause your speaker to play incorrect notes for C, G, and A and may even mean that your speaker will not make any sound at all for some of the notes that it should play.

The example above on the right shows an example where the while loop variable i is initialized incorrectly. Its initial value should be 0, but the code sets it to 1. (Remember computer scientists always begin counting from 0 instead of 1.) This means that the loop will check the last six keys, but it will never check the first key or play the first note. That is, the program will never execute checkPianoKey(keys[0], notes[0]);. This means that the first key on the piano will not work.

FIX INCORRECT VARIABLE INITIALIZATIONS

If you are having problems with a particular part of your program make sure you investigate the variables that relate to that part of the code in your troubleshooting. Many different problems can be caused by incorrect variable initializations, so it is an important thing to check for. If you find an incorrect variable initialization, correct the problem and recompile and upload your code. Check your project's behavior to see if your edit has fixed the problem.

PROBLEMS WITH DELAY

error example:

```
void loop() {
    digitalWrite(led, HIGH);
    delay(1);
    digitalWrite(led, LOW);
    delay(1);
}
```

fixed example:

```
void loop() {
    digitalWrite(led, HIGH);
    delay(1000);
    digitalWrite(led, LOW);
    delay(1000);
}
```

error example:

```
void song() {
    tone(speaker, E);
    delay(2000);
    tone(speaker, D);
    delay(2000);
    tone(speaker, C);
    delay(2000);
    noTone(speaker);
    delay(20000);
}
```

fixed example:

```
void song() {
    tone(speaker, E);
    delay(2000);
    tone(speaker, D);
    delay(2000);
    tone(speaker, C);
    delay(2000);
    noTone(speaker);
    delay(2000);
}
```

symptoms

Misuse of delay statements can cause a few different problems. Problem delays can be too short, too long, or missing altogether. If a delay is too short, it may seem like things that should be happening are not happening. For example, in the code above on the left, the delays after the digitalWrite(led, HIGH); and digitalWrite(led, LOW); statements are very short. The statement delay(1); tells your project to stop and wait for one millisecond—that's 1/1000 of a second, a tiny amount of time. In this program, the LED will turn on and off so quickly that you will not be able to see it blinking. Instead, your LED will look like it's on dimly all of the time. The same kind of thing can happen with speakers. If you have a very short delay after a tone statement, you may not hear anything or may only hear a faint clicking sound.

If your program seems to never respond to you or to only respond very slowly—if your monster or piano never responds to your touch or responds very slowly—you may have a delay that is too long or a series of delays that are taking too much time. For example, in the code above on the right, the last line of the song procedure (delay(20000);) will make your project stop and do nothing for 20,000 milliseconds—that's 20 seconds, a very long time! This means you'll have to wait 20 seconds before your project will be able to detect your touch, blink, or make sounds again. This is an extreme example. Your behavior may be much more subtle. For example, in the monster project, if your blinkPattern procedure takes a few seconds to run (say three seconds) then you may have to hold onto your monster's paws for three seconds before it responds to your touch. This may mean that your monster is less responsive than you'd like it to be.

Remember as you think about the delays in your code, that programs execute line by line in order and the LilyPad can only do one thing at a time. It needs to finish a delay before it moves on to the next part of your program.

Remember also that if your project is sending data back to your computer with Serial.println statements, you will overload your computer with too much data if your program does not have a delay statement. If this happens, your computer will probably crash.

FIX PROBLEMS WITH DELAY

If you are having problems like any of the ones described above, read through your program line by line, paying particular attention to the delay statements. If a component like an LED or a speaker seems like its not working, look for delays that are too short and, if you find any, make them longer. If a sensor seems like it's not working, look for delays that are too long or entire sections of code that take a long time to execute. Try to shorten the delays or edit sections of code so that they execute more quickly. If your computer has crashed from a data overload, add a delay statement to your program right after your Serial.println statement.

Recompile and upload your code. Check your project's behavior. You may need to adjust the delays in your program a few times before you get a behavior you're happy with.

MATERIALS REFERENCE

ELECTRONIC MATERIALS & TOOLS

Coin cell battery

A 3-volt battery whose code name is CR2032. This code means that the battery is 20mm in diameter and 3.2mm thick. Flat, round batteries like these are called coin or button cell batteries because of their shape.
Where to buy? Online retailer like Amazon. Sparkfun sells this battery, product #338, but its slightly more expensive.
Approximate cost: $0.50

LilyPad coin cell battery holder

A sewable battery holder for the CR2032 coin cell battery.
Where to buy? https://www.sparkfun.com/products/11285
Approximate cost: $6.00
Alternate options: If you want to make your own battery holder out of fabric you can find a tutorial here: http://www.kobakant.at/DIY/?p=52

Conductive thread

A thread capable of carrying electric current. The thread referred to in this book is a stainless steel thread spun from fine stainless steel wires. Other types of conductive thread include silver-plated threads and gold-wrapped embroidery threads.
Where to buy? https://www.sparkfun.com/products/10867
Approximate cost: $3.00
Alternate options: You can purchase silver-plated thread from plug and wear: http://www.plugandwear.com/

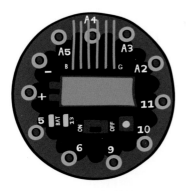

Lilypad Arduino SimpleSnap

A small snap-on computer. This programmable LilyPad Arduino board contains an ATmega328 microcontroller and a built-in rechargeable battery. For more information see: http://lilypadarduino.org/?p=289
Where to buy? https://www.sparkfun.com/products/10941
Approximate cost: $30.00
Alternate options: You can use any other LilyPad Arduino board to replace the SimpleSnap and Protoboard combination. For more information see: http://lilypadarduino.org/?cat=4

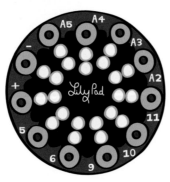

Lilypad Arduino Protoboard

A sewable board with a ring of male snaps on its outer edge that mates with the LilyPad Arduino SimpleSnap. Sew this board to your project to be able to snap the LilyPad Arduino SimpleSnap on and off of it.
Where to buy? https://www.sparkfun.com/products/10940
Approximate cost: $10.00
Alternate options: You can replace the protoboard with sew-on snaps (size 1/0) or riveting snaps (size 10). For more information see: http://lilypadarduino.org/?p=289 If you use a LilyPad that is not the SimpleSnap, you do not need a Protoboard at all.

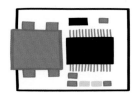

FTDI Breakout Board / Programming board

The board that connects a LilyPad Arduino board to a computer so that it can be programmed.

Where to buy? https://www.sparkfun.com/products/10275

Approximate cost: $15.00

Alternate options: You can replace the FTDI Breakout Board and USB cable with an integrated FTDI cable: https://www.sparkfun.com/products/9718

Lilypad Speaker / Lilypad Buzzer:

A small sewable speaker.

Where to buy? https://www.sparkfun.com/products/08463

Approximate cost: $8.00

Lilypad LED

A small sewable LED (light-emitting diode).

Where to buy? https://www.sparkfun.com/products/

Approximate cost: $1.00

Alternate options: You can replace LilyPad LEDs with standard "through hole" LEDs. Simply twist their legs into sewable loops like so:

You can purchase through hole LEDs from your local Radioshack or sparkfun: https://www.sparkfun.com/products/9590

Mini-USB Cable

A USB 2.0 A to Mini-B Cable. The cable that connects the FTDI Breakout Board / Programming board to your computer. This type of cable is used to connect many cameras and cell phones to computers and chargers. Check to see if you already own one before you buy a new one.

Where to buy? Online retailer like Amazon. Sparkfun sells a red USB cable, product #598.

Approximate cost: $5.00

Alternate options: You can replace the FTDI Breakout Board and USB cable with an integrated FTDI cable: https://www.sparkfun.com/products/9718

LilyTiny

A tiny sewable computer that is programmed to control LEDs. This board is powered by an ATtiny85 microcontroller. For more information see: http://lilypadarduino.org/?p=523

Where to buy? https://www.sparkfun.com/products/10899

Approximate cost: $7.00

MATERIALS REFERENCE

CRAFT MATERIALS & TOOLS

Embroidery Thread or "Floss"

A special heavy-weight thread used for embroidery.
Where to buy? Your local craft store, or an online retailer like Amazon.
Approximate cost: $1.00

Glue

Any standard craft glue will work well for sealing knots, including fabric glue or Elmers® glue. Use a fabric glue or Iron-On adhesive to attach one piece of fabric to another.
Where to buy? Your local craft store, or an online retailer like Amazon.
Approximate cost: $5.00

Large-eyed needle

"Chenille" needles sized 18-24 are very easy to thread and fit through the holes in LilyPad pieces. If you are a more experienced sewer, you may want to use a smaller needle, but be careful, smaller needles are much harder to thread.
Where to buy? https://www.sparkfun.com/products/10405
Approximate cost: $2.00

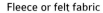

Polyester filling

The material that's inside pillows and stuffed animals. The soft squishy center of your stuffed monster.
Where to buy? Your local craft store, or an online retailer like Amazon.
Approximate cost: $15.00

Fleece or felt fabric

These thick non-stretch fabrics are easy to sew.
Where to buy? Your local craft store, or an online retailer like Amazon.
Approximate cost: $10.00 per yard

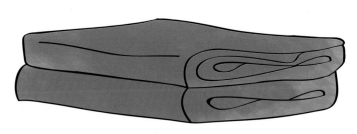

Aluminum foil

Standard aluminum foil. Do not try to use the non-stick variety.
Where to buy? Your local grocery store.
Approximate cost: $2.00

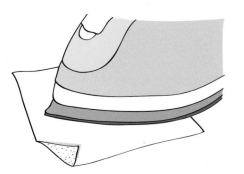

Iron-on adhesive

Thermoweb®'s Ultra Hold Heat-n-Bond adhesive. You'll need the sheet, not the tape.
Where to buy? Your local craft store, or an online retailer like Amazon.
Approximate cost: $15.00

Miscellaneous craft materials

Scissors, paper, pencils, chalk, etc.
Where to buy? Your local craft store, or an online retailer like Amazon.
Approximate cost: $5-10.00

abstraction: the process of creating pieces of code such that a programmer does not need to understand how the code works to use it in her program. You do not necessarily need to know how a piece of code, like a procedure, works to use it in a program. As long as you know the procedure's name, its inputs, its outputs, and what it does, you can use it. For example, in the piano tutorial, when you use the readCapacitivePin procedure, you're taking advantage of abstraction (see page 116).

analog: in electronics, a signal that conveys information through a continuously changing value. For example, the sensor you constructed in the monster tutorial is an example of an analog sensor since it conveys information through a continuously changing value—electrical resistance. Analog signals are different from digital signals, which convey information through discrete changes in values. See also **digital**.

array: a list or table in code. Arrays provide a way to store an ordered collection of data in your program. For example, the statement int keys[7] = { key1, key2, key3, key4, key5, key6, key7 }; declares an array called keys that consists of seven entries, key1, key2, key3, key4, key5, key6, key7, of type int. The entries in an array are accessed using numbers in square brackets. The access numbers begin with zero. For example, the first entry in the keys array can be accessed with the statement keys[0], the second entry with the statement keys[1], and so on. See page 139 for more information.

baud rate/baud: See **communication speed**.

C (programming language): a general-purpose programming language developed in the late 1960s by Dennis Ritchie, a researcher at AT&T Bell Labs. All Arduino programs are written in C. However, the Arduino environment has many features that aren't part of standard C, including built-in procedures like digitalWrite and delay, so the Arduino language can be said to be a dialect (or version) of C. C is one of the most widely used programming languages in the world.

call (a procedure): calling a procedure means that you use the procedure in the main part of your program. For example, you call Arduino's built-in delay procedure when you include the statement delay(1000); in the loop part of your program.

capacitance: a material's ability to store electric charge.

capacitive sensor: a sensor based on changing capacitance. That is, a sensor that changes its capacitance in response to a stimulus, often touch. For example, each piano key in the piano tutorial is a capacitive sensor because the capacitance of a key changes when a person touches it.

circuit: a network of electrical devices. It is generally a closed loop that includes a power supply and other electrical elements like LEDs and switches. The loop structure enables the flow of electric current from (+) to (-).

code or **source code**: a collection of instructions that will be carried out by a computer (or LilyPad) and that are written in a programming language. Any piece of an Arduino program can be called code. See also **program**.

comment: a piece of text in a program that is ignored by the compiler and the computer (or LilyPad) that executes the code. In the C programming language there are two kinds of comments: comments that extend for several lines (like paragraphs) and comments that are written on a single line. To create a comment that spans multiple lines, begin the comment with /* and end it with */. To create a comment that is only on one line, begin it with //. Comments are shown in a greyish brown in the Arduino environment. For example, the line // set all keys to be inputs is a comment. So is the line // the loop routine runs over and over again forever. See page 52 for more information.

comment out (code): to turn a piece of code into a comment. Commenting out a piece of code allows you to temporarily prevent that code from executing—since comments are ignored by the compiler—while keeping the text of the code in your program. To comment out a single line of code, add two slash characters // to the beginning of the line. To comment out a larger piece of code, add /* characters to the beginning of the section you want to comment out and */ to the end of the section. You should see the commented out section turn grey in the Arduino window. When you want to include the code in your program again, un-comment it; change it from a comment back into code by removing the // or /* and */ characters. For example, the statement delay(1000); is commented out when two slash characters are placed in front of it like so: //delay(1000); See also **comment**.

communication speed: (also known as **baud rate**) the speed at which one computational or electronic device communicates with another. Communication speed is measured in bits per second. 9600 bits per second (or "baud") is the standard communication speed for most Arduino applications. Before sending data from a LilyPad to a computer in an Arduino program, you must specify the communication speed using the Serial.begin procedure. For example, the statement Serial.begin (9600); sets the LilyPad's communication speed to 9600 pits per second. See also **serial port**.

compile: to translate a program, written in a programming language like C, into a machine-readable code like hex code. In the Arduino environment, clicking on the check mark icon in the toolbar compiles C code that you wrote into hex code that a LilyPad can understand.

compile error: an error in a program that is detected when software attempts to compile the program. Compile errors are often syntax errors like misspellings or missing semi-colons. For example, since the word "delay" is misspelled in the statement dellay(1000);, code containing this statement will generate a compile error. See page 155 for more information on compile errors.

condition: a statement that is always either true or false. Conditions are often comparisons using less than (<), greater than (>), and equal to (==) operators. For example, the condition x < 10 is true when x is less than 10 and false when x is 10 or greater. Conditions are an important part of conditional statements, where they appear in parentheses. For example, in the line if (i<10), (i<10) is the condition. See also **conditional statement**.

conditional statement: a block of code that does one thing if a condition is true and another thing if the condition is false. if else statements and while loops are examples of conditional statements. See pages 95 and 143 for more information. See also **condition**.

conductive: a conductive material is one that an electrical current flows through easily. Metals like copper, silver, and aluminum are highly conductive. Conductive materials in the projects in this book include the tabs on the LilyPad components and the conductive (stainless steel) thread used to sew these components together. The opposite of a conductive material is a insulating material. Insulating materials—like plastic, glass, and fabric—resist the flow of electricity. Electrical current does not flow through insulating materials.

constant: a named variable in a program whose value never changes. Constants are declared like variables with a type and a name. However, constants require an additional code word, const, at the beginning of the statement. For example, the numberOfKeys constant in the piano project is declared with the statement const int numberOfKeys=7;. This constant stays the same throughout the Piano program.

digital: information represented using discrete values. In computing, information represented using 1s and 0s. In electronics, a signal that conveys information through discrete values (HIGH/LOW) instead of continuously varying values. For example a switch is an example of a digital sensor since it conveys information in a discrete manner; a switch is either open or closed. Digital signals are different from analog signals, which convey information through continuous changes in values. See also **analog**.

electric current: the flow of electricity through a circuit. Current travels from the (+) side of a battery (power), through the components connected in the circuit, and back to the (-) side of the battery (ground). Electric current only flows through a circuit if the path is complete, that is, if there are no breaks in the circuit. Electric current is measured in amps.

electronic textiles or **e-textiles**: fabrics that include soft electrical circuitry and embedded electronics like sensors, lights, motors, and small computers. Designers of e-textiles strive to keep things soft by using new materials like conductive thread, conductive fabric, and flexible circuit boards.

energy: electrical energy is the ability of a power supply to run a circuit over time, to light up an LED, or make sounds with a speaker. The amount of energy stored in a battery is equal to its amp-hour rating multiplied by its voltage rating. Energy is measured in watt-hours (Wh). For example, a 3-volt battery with an amp-hour rating of .25 amp-hours stores .75 watt-hours of energy.

execute (a program): (also **run**) the act of carrying out the instructions specified by a program. A computer—like the LilyPad Arduino—executes programs line-by-line in the exact order in which they are written.

frequency: the speed or pitch of a sound wave. Sound is created by vibrations of molecules in the air. When the molecules vibrate very quickly—at a high frequency—you hear a high note; when they vibrate more slowly—at a low frequency—you hear a low note. Frequency is measure in pulses per second or Hertz (Hz).

ground (-): the negative terminal of a battery or other power supply in a circuit; 0 volts. Also, any part of a circuit that is at 0 volts. Ground is the reference point in a circuit from which all other voltages are measured. The color black is used to denote ground in circuit diagrams and drawings. In Arduino code, ground is referred to as LOW. See also **low** and **power**.

hex code: short for hexadecimal code. A numerical code based on 16 symbols (a base 16 number system). When Arduino code is compiled it is translated into hex code. This code is then loaded onto the LilyPad Arduino, which can read and execute it.

high or HIGH: in Arduino code, the term used to refer to power (+) in electrical circuits. Power, (+), and HIGH refer to the positive terminal of a battery or other power supply in a circuit. See also **power** and **low**.

initialize (a variable): a variable is initialized when a value is assigned to it for the first time. For example, the statement int Led = 13; declares a variable called Led and initializes it to 13. The statement assigns the value of 13 to the variable Led.

input (device): an electrical component that gathers information from the world. This information could include how hard a sensor is being pressed, the current temperature, or the ambient sound level is. Input devices include switches, touch sensors, thermometers, cameras, and microphones. All sensors are inputs. In Arduino, information is collected from inputs with digitalRead and analogRead statements. The readCapacitivePin procedure used in the piano tutorial also collects information from an input (a touch sensor). See also **read** and **output** (device).

input (to a procedure) or **input variable**: information required by a procedure. In this book, usually a number supplied in parentheses after the procedure name in a procedure call. For example, in the statement delay(1000);, the number 1000 is the input. Input variables help programmers write procedures that are applicable to a wide range of situations. Inputs enable you to carry out the same basic set of instructions or statements, but with different values. For example, the fact that the built-in procedure delay has an input, lets you delay for different amounts of time in your programs. When you call delay with an input of 1000, your program pauses for one second. When you call delay with an input of 100, it pauses for 1/10 of a second, and so on. Similarly, in the monster tutorial, the duration input variable to the song procedure lets you play the song you created at different speeds when different values are provided as inputs (see page 86).

int or int: a data type used in Arduino programs used to declare integer, or whole number, variables. Almost all of the variables and numbers used in the programs in this book are of type int. For example, the statement int led = 13; declares a variable called led of type int. See also **type** and **integer**.

integer: a whole number. A number that does not have a fractional or decimal component. 1, 2, 5, 20, 32, 548870, and 1000 are examples of integers. 1.2, 50.001, ¾, and .1 are examples of numbers that are not integers.

LED: short for light-emitting diode. LEDs contain an electroluminescent material—a material that glows when electrical current flows through it. LEDs are polarized. That is, they have a (+) and a (-) side and will only light up when current flows from their (+) to (-) side. If you attach an LED backwards in your circuit, it will not work. LEDs are more efficient than most other light sources. That is, they produce more light with less energy than most other types of lights. See also **polarity**.

logical error: an error that occurs when your code compiles and uploads, but doesn't behave the way you want it to. These errors are the trickiest to find and fix because the computer doesn't give you any feedback about what might be causing problems, like it does with compile and upload errors. See page 160 for more information.

low or LOW: in Arduino code, the term used to refer to ground (-) in electrical circuits. Ground, (-), and LOW refer to the negative terminal of a battery or other power supply in a circuit. LOW is always 0 volts. See also **ground** (-) and **high**.

memory: where an Arduino program is stored once it's uploaded to a LilyPad. Once a program is stored in the LilyPad's memory, the LilyPad can run the program independently of the computer.

microcontroller: a small computer chip that stores and execute programs and controls electronics. Microcontrollers, like most computers, have a memory that is used to store programs and program data, and a processor that is used to interpret and execute programs. Microcontrollers also have pins that can be used to control input and output devices. When input and output devices like sensors and LEDs are attached to the pins, the microcontroller can read electrical signals from the inputs and send electrical signals out to the outputs. The LilyTiny board contains an ATtiny85 microcontroller and the LilyPad Arduino SimpleSnap board contains an ATmega328 microcontroller. See also **pin**.

milliseconds: 1/1000 of a second. 1000 milliseconds (1000 ms) is one second. The input to the Arduino procedure delay is in milliseconds. Thus, delay(1000); pauses program execution for one second, delay(100); pauses program execution for 1/10 of a second, and so on.

open source: a term used to describe a program whose source code is publicly available for you to use, examine, and modify. The sharing of open source software, also known as free software—free as in speech, not free as in beer—can enable people to collaborate on projects by helping them extend and expand on each other's work. You are making use of open source software and open source hardware throughout this book. The Arduino software is open source and the designs of the LilyPad boards are also open source. You use open source software most directly in the Piano tutorial when you use the readCapacitivePin procedure—code that was written by someone else and published in open source fashion online.

output (device): an electrical component that takes action—does something—in the world. Actions could include lighting up, moving, making sound, or changing shape. Output devices include lights, motors, speakers, and display screens. In Arduino, outputs are controlled by digitalWrite statements. See also **write** and **input** (device).

parallel circuit: components in a parallel circuit have all of their (+) sides connected together and all of their (-) sides connected together. This configuration allows for all of the components to receive the same voltage.

pin: part of a microcontroller that can attach to and control an input or output device. Each microcontroller pin can control either an input device, like a switch, or an output device, like an LED. The pins on a microcontroller look like tiny legs coming out of the controller's black square body. On the LilyTiny and LilyPad Arduino SimpleSnap boards, microcontroller pins are connected to sewable tabs and snaps. When an input or output device is attached to a tab or snap, the microcontroller can control that component. This book uses the terms pin and tab interchangeably. See also **tab** and **microcontroller**.

polarity: electrical orientation or direction. Electricity flows in one direction in a circuit, from (+) to (-). Due to this property, electrical circuits are said to be polarized. Electrical components with polarity will only work properly when electrical current flows through them in a particular direction, from their (+) to (-) side. LEDs are examples of components with polarity.

power (+): the positive terminal of a battery or other power supply in a circuit. Generally the highest possible voltage in a circuit. The color red is used to denote power in circuit diagrams and drawings. Note: Different circuits may have different power (+) voltages. For example, in a circuit that uses a 3-volt battery, power is +3 volts. In a circuit that uses a 3.7 volt battery, power is +3.7 volts. In Arduino code, power is referred to as HIGH. See also **high** and **ground**.

procedure: a block of code that is given a unique name. A procedure may have one or more inputs and it may return a value. When a procedure is called, the program jumps to the place in the program where the procedure is defined, executes the block of code that makes up the body of the procedure, and then jumps back to the point right after the procedure was called in the program. The Arduino language has a library of built-in procedures like delay, digitalWrite, and analogRead. You can also define your own procedures (see for example, page 76). See also **input** (to a procedure) and **return**.

program: a set of instructions to be carried out by a computer (or LilyPad). A program is written in a programming language. Programs can also be called pieces of code. A program does its work when a computer runs or executes its instructions by following them in order. See also **code**.

programming language: a language that enables people to write instructions for computers. Programming languages generally have limited vocabularies and very strict formatting rules. This enables them to be read and understood by machines. There are many different programming languages. You may have heard of languages like Python, Java, C++, and Scheme. Arduino programs are written in the C programming language.

read (from a pin): the act of gathering information from an input device. Information is collected via an input device attached to a pin on a LilyPad (or other microcontroller). In Arduino, information is read from a pin with the statements digitalRead and analogRead. digitalRead(pin); tells you whether the pin is HIGH or LOW (see page 95). analogRead(pin); measures the voltage level of the pin and gives you a number between 0 and 1023 that corresponds to the voltage (see page 96). See also **write** and **input** (device).

resistive sensor: a sensor based on changing resistance. That is, a sensor that changes its electrical resistance in response to a stimulus. For example, the touch sensor in the monster tutorial is a resistive sensor because the electrical resistance between the monster's two paws changes when a person touches and squeezes them.

return: a procedure is said to return when it finishes executing. Some procedures return a value. That is, the procedure finishes executing and provides the results of its execution back to the program that called it. For example, the procedure analogRead(pin); returns the value it read from the input pin. See also **procedure**.

run: see execute.

running stitch: the most basic stitch in hand sewing. Also called a straight stitch. This stitch is created by passing a needle and thread up and down through a piece of fabric along a line (see page 13). A good running stitch consists of neat, even stitches of about ¼″ (6mm) in length.

sensor: an electrical component that gathers information from the world, for example: how hard a touch sensor is being pressed, what the current temperature is, or what the ambient sound level is. Sensors include touch sensors, thermometers, cameras, and microphones. All sensors are inputs. In Arduino, information is collected from sensors with the analogRead statement. The readCapacitivePin procedure used in the piano tutorial also collects information from a sensor. See also **read** and **input** (device).

serial port: the communication channel through which a computer communicates with a LilyPad and vice versa. The serial port connection—the USB attachment between the LilyPad and your computer—allows Arduino to upload programs to your LilyPad and allows your LilyPad to send information back to your computer with Serial.println and Serial.print statements. See also **communication speed**.

short circuit or **short**: the direct electrical connection of a power supply's (+) and (-) sides. When a short circuit occurs, the circuit's power supply releases a tremendous burst of energy. A short circuit can ruin a battery and can electrocute or burn you if the power supply is powerful enough. The batteries you're using for the projects in this book won't shock or burn you even if you create a short circuit. But, if you do create one—by connecting your project's (+) and (-) sides or by connecting other traces that should not be touching, like the trace connecting the (+) side of your LED and the trace connecting the (+) side of your speaker in the monster project—your project may not work and you may quickly ruin your battery. See pages 19 and 152 for more information.

statements: computer sentences; lines of code written in a programming language that tell a computer to do something. Simple statements in Arduino end with a semicolon the way that English sentences end with a period. The line digitalWrite(led, HIGH); is a simple statement. So are the lines delay(1000); and song(2000);. More complex statements can span several lines. Examples of complex statements include if statements and while loops.

switch: a circuit component that is always in one of two states: open (disconnected) or closed (connected). In a simple circuit, like the one in the bookmark tutorial (see page 20), the flow of electricity through a circuit is stopped when the switch is open and restored when the switch is closed.

syntax: the rules that define the structure of a programming language. These are the spelling, punctuation, capitalization, and formatting rules you must follow when you're writing a program. Different programming languages have different syntaxes.

syntax error: an error in a program that happens when you do not follow the programming language's syntax. See also **syntax** and **compile error**.

tabs: the silver-rimmed holes in the sewable LilyPad boards used in the projects described in this book. On the LilyTiny and LilyPad Arduino SimpleSnap boards, tabs are connected to microcontroller pins. See also **pin**.

threshold: a cutoff value used in a program such that one behavior happens if a variable has a value below the threshold and a different behavior happens in the variable has a value above the threshold (see for example page 100). Thresholds are often used in conditional statements of the form if (variable < threshold) or if (variable > threshold).

trace: a conductive connection between two components in a circuit.

type: computer programs handle different kinds of information including whole numbers, decimal numbers, text, and images. Before a program can manipulate a piece of data, it needs to know what type of data it is dealing with—whether it is dealing with an image or a piece of text, for example. In most programming languages, each variable used in a program must have a specified type when it is first declared. In the projects described in this book, almost all of the variables you use are of type integer or int. They are whole numbers. In Arduino, integer variables are declared with a statement of the form int variableName; or int variableName=#;. In these statements, int specifies the variables' type. See also **variable declaration**, **int** and **integer**.

upload: the act of sending code that has been compiled and converted into hex code from a computer to a LilyPad Arduino.

variable: a named location in a program's memory that can store values. Variables enable you to give names to information and elements used in your program. They make programs easier to write, understand, and modify. Variables are generally listed at the beginning of an Arduino program. This list as analogous to the list of ingredients at the beginning of a cooking recipe. Every Arduino variable has a type that is specified when it is declared. See also **variable declaration**, **variable initialization** and **type**.

variable declaration: a statement introducing (or "declaring") a variable and its type. For instance, the statement int myVariable; declares a variable named myVariable of type int, short for integer. All variables in a program must be declared before they can be used. Variables are often declared and initialized in a single statement. The statement int myVariable=13;, for example, both declares a variable called myVariable and initializes it to the value 13. See also **variable**, **variable initialization** and **type**.

variable initialization: a statement that gives a variable a value for the first time. For example, the statement int myVariable=13; declares a variable called myVariable and initializes it to the value 13. Variables are often declared and initialized in the same statement, but not always. The statement int myVariable=13; can be broken down into two separate statements. The statement int myVariable; declares the variable myVariable. The statement myVariable=13; initializes myVariable, giving it its initial value. See also **variable** and **variable declaration**.

write (a pin): the act of controlling an output device attached to a pin on a LilyPad (or other microcontroller) by sending it electrical signals. In Arduino, electrical signals are written to a pin with the statement digitalWrite. digitalWrite(pin, value); sets the pin to be either HIGH (maximum circuit voltage, power (+)) or LOW (minimum circuit voltage, 0 volts, ground(-)). See page 74 for more information about digitalWrite. See also **read** and **output** (device).

INDEX

Abstraction: 116, **172**

Amp-hours: 18, 19, 174

Analog: 96, **172**

analogRead: 96, **97-98**, 174, 176

Arduino (software): 2, **44-46**, **47-49**, **52-55**, **56-58**, 60, 73, 74, 82, 83, 86, 88, 94, 96, 99, 114, 115, 116, 127, 132, 136, 137, 147, 148, 155, 156, 157, 158, 159, 160, 172, 173, 174, 175, 176, 177, 178, see also **LilyPad Arduino**

Array: **139-142**, 146, **172**

ATmega328: 43, 45, 168, 175

Battery/coin cell battery/3-volt battery: 6, 8, 9, 10, 16, 17, 18, 19, 20, 21, 25, 26, 28, 31, 32, 33, 34, 35, 36, 38, 43, 44, 69, 74, 79, 90, 105, 123, 149, 151, 154, **168**, 174, 175, 176, 177

Battery holder/coin cell battery holder: 6, 8, 9, 10, 11, 12, 13, 14, 16, 20, 21, 26, 30, 31, 32, 33, 34, 36, **168**

Baud/baud rate: 99, 125, **172**, 173, see also **Communication speed**

Board: see **Circuit board**

C (programming language): 47, 48, 52, 56, **172**, 173, 176

Call/calling (a procedure): 76, 77, 84, 85, 86, 88, 95, 116, 129, 130, 139, 142, 143, 144, **172**, 175

Capacitance: 116, 118, **172**

Capacitive sensor: see **Sensor, capacitive**

Circuit: 8-9, 16, **18-20**, 22, 23, 35, 36, 68-69, 93,111,151, 160, 162, 165, **172**, 174, 175, 176, 177, 178, see also **Short circuit**

Circuit board/board: 1, 20, 43, 174

Code/source code: v, 2, 37, 46, 47, 48, 49, 51, 52-55, 56, 57, 59, 60, 61, 73, 74, 75, 76, 77,78, 82, 83, 84, 85, 86, 87, 88, 89, 94, 95, 96, 97, 99, 100, 101, 114, 115, 116, 117, 120, 121, 122, 127, 128, 129, 130, 131, 133, 137, 138, 139-145, 147, 148, **172**, 173, 174, 175, 176, 177, 178, see also **Program/programming** and **Troubleshooting solutions, code**

Comment: 47, 51, **52**, 53, 54, 56, 73, 88, 114, 120, 121, 159, **173**

Comment out: **52**, 84, 95, 97, **173**

Communication speed: 99, 114, 172, 173, 177, see also **Baud/baud rate**

Compile: 46, 47, **48**, 51, 52, 57, **173**, 177

Compile error: 46, 48-49, 54-55, 57, 58, **59**, 61, 80, 91, 94, 103, 115, 124, 135, 146, **155-159**, 164, **173**, 177, see also **Syntax errors** and **Troubleshooting solutions**

Condition: 95, 103, 125, 143, 160, 163, 164, **173**

Conditional statement: 95, 103, 120, 121, 127, 163, 164, **173**, 177

Conductive: 1, 5, 6, 8, 9, 10, 18, 19, 20, 23, 26, 28, 32, 64, 70, 71, 93, 96, 98, 108, 110, 111, 112, 113, 118, 123, 126, 151, 154, 168, **173**, 174, 178

const/**constant**: 141, **173**

Debug/debugging: 49, 78

Declare (a variable): 53, 56, 158, 172, 173, 174, 175, **178** , see also **Variable declaration**

delay: 51, 54, 61, 75, 76, 84, 85, 86, 88, 99, 124, 125, 131, 132, 135, 160, 164, 166, 172, 173, 175, 176, 177

Design Instructions:
    Bookmark: 8-9
    Bracelet: 28-30, 32
    Circuit: 9, 31, 68-69, 111
    Monster: 66-67
    Piano: 110-111
    Capacitive sensor: 110
    Resistive sensor: 92-93

Digital: 96, 172, **174**

digitalRead: **95**, 96, 97, 98, 174, 176

digitalWrite: 51, 52, 54, 55, 61, **74**, 75, 76, 78, 95, 161, 166, 172, 176, 177, 178

Electric current: **18**, 20, 168, 172, **174**

Electrocute/shock: 20, 177

Electronic circuit: 5, See also **circuit**

Electronic textiles/e-textiles: v, **1**, 2, 3, 5, 41, **174**

Energy: 18, 19, 20, **174**, 175, 177

Energy ratings (of batteries): 19

Execute (a program): 46, 51, 53, 56, 74, 88, 138, 164, 165, 166, 173, **174**, 175, 176, 177, see also **Run**

Frequency: 82, 142, 147, **174**

FTDI board/FTDI breakout board: 42, 43, 44, 50, 60, 64, 73, 108, 114, **169**

Ground (-): 19, 20, 74, 95, 96, 98, 152, **174**, 175, 176, 178, see also LOW

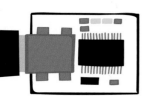

FTDI board

LED

H

Hertz: 82, **174**

Hex code: **48**, 49, 173, **174**, 178

HIGH: 19, 51, 54, 55, 74-75, 95, 96, 98, 102, 103, 161, 173, **174**, 175, 176, 178, see also **Power (+)**

I

if else: **95**, 100, 121, 163, 173

Initialize (a variable): 53, 61, 80, 91, 94, 96, 103, 114, 121, 124, 127, 155, 161, 165, **174**, 178

Input (device): **74**, 92, 94, 95, 99, 127, **174**, 175, 176, 177, see also **Sensor**

Input (to a procedure)/Input variable: 74, 75, 76, 77, 82, 83, 85, **86-87**, 98, 115, 116, 117, **129-131**, 139, 141, 142, 143, 172, **175**, 176

int/integer: 53, 54, 55, 73, 83, 86, 94, 116, 120, 129, 139, 140, 141, 172, 173, 174, **175**, 178

Internet: v, 42, 46, 116

L

LED/light-emitting diode: 1, 2, 5, 6, **8**, 9, 10, 13, 14, 15, 16, 17, 18, 19, 22, 23, 25, 26, 28, 30, 31, 32, 33, 34, 35, 36, 37, 39, 41, 43, 50, 51, 52, 53, 54, 55, 58, 61, 64, 66, 68, 69, 70, 71, 72, 73, 74, 75, 79-80, 81, 84, 88, 93, 94, 95, 100, 151, 152, 154, 161, 162, 166, **169**, 172, 174, **175**, 176

LilyPad Arduino/LilyPad Arduino SimpleSnap/ LilyPad: v, 1, 2, 41, 42, **43**, 44, 45, 46, 47, 49, 50, 51, 52, 53, 54, 55, 58, 60, 61, 64, 66, 68, 69, 73, 74, 75, 79, 81, 82, 83, 84, 90, 92, 94, 96, 97, 98, 99, 100, 105, 107, 108, 110, 111, 114, 115, 117, 119, 120, 121, 123, 126, 127, 130, 131, 132, 137, 138, 141, 142, 144, 146, 148, 149, 160, 163, 164, 166, **168**, 170, 172, 173, 174, 175, 176, 177, 178

LilyPad Arduino Protoboard/Protoboard: 64, 66, 68, 70, 71, 72, 73, 74, 81, 83, 93, 94, 97, 108, 112, 113, 114, 119, 126, 151, 152, 154, **168**

LilyPad Buzzer: see **LilyPad Speaker**

LilyPad coin cell battery holder: **168**, see also **Coin cell battery holder**

LilyPad LED: 6, 26, 51, 64, **169**, see also **LED**

LilyPad Speaker/Buzzer: 64, **66**, 68, 69, 74, 81, 82, 83, 84, 90, 95, 100, 107, 108, 110, 111, 119, 120, 121, 123, 124, 147, 148, 151, 152, 161, 162, 164, 165, 166, **169**, 174,176

LilyTiny: 1, 26, **28**, 31, 32, 33, 36, 39, 43, 151, 152, 154, **169**, 175, 176, 177

Logical errors: **61**, 80, 91, 103, 124, 125, 135, 146, **160-166**, see also **Troubleshooting solutions**

L

loop/loop section (in Arduino programs): 55, 56, 57, **58**, 74, 75, 76, 84, 85, 86, 87, 88, 89, 95, 96, 97, 99, 101, 114, 117, 121, 127, 129, 130, 131, 132, 135, 138, 139, 142, 143, 144, 146, 151, 163, 164, 165, 172, 173, 177

LOW: 19, 55, 74, 75, 95, 96, 98, 102, 174, **175**, 176, 178, see also **Ground (-)**

M

Memory: 49, 50, 53, **175**, 178

Microcontroller: 28, 43, 53, 74, 168, 169, **175**, 176, 177, 178

Milliseconds: 51, 75, 83, 86, 121, 166, **175**

Mini-USB cable/USB cable: 42, 44, 45, 64, 73, 84, 96, 108, 114, **169**, 177

N

noTone: 83, 84, 120, 121, 124, 125, 132

O

Open source: 115, 175

Output (device): 57, 74, 84, 94, 95, 115, 116, 118, 121, 158, 161, 172, 174, 175, 176, 178

P

Parallel/parallel circuit: **22**, 39, 176

Pin: 17, **28**, 35, 36, **43**, 53, 61, 68, 69, 72, 73, 74, 75, 79, 80, 81, 82, 83, 84, 90, 91, 93, 94, 95, 96, 98, 102, 110, 112, 113, 116, 117, 118, 119, 120, 121, 123, 124, 125, 126, 134, 140, 152, 162, 165, 175, **176**, 177, 178

pinMode: 56, 57, **74**, 84, 94, 127, 141, 161

Polarity: 9, 17, 35, 39, 79, 90, 151, 154, 175, **176**

Power (+): 19, 20, 74, 95, 98, 152, 174, **176**, 178, see also HIGH

Procedure: 55, 74, 75, **76-77**, 80, 82-83, 84, 85-86, **86-87**, 88, 89, 91, 95, 96, 98, 99, 100, 101, 115, 116, 117, 118, 120, 121, **129-131**, 132, 136, 137, 138, 139, 143, 147, 148, 158, 166, 172, 173, 174, 175, **176**, 177

Program/programming: v, 1, 2, 28, **36-37**, 41, 42, 43, 44, **46**, 63, 73, 74, 75, 76, 77, 78, 83, 84, 85, 88-89, 94-101, 107, 110, 114, 115, 116-117, 127-133, 136, 140, 141, 142, 143, 144, 145, 147, 148, 155, 156, 157, 158, 159, 160, 161, 163, 164, 165, 166, 168, 169, 172, 173, 174, 175, **176**, 177, 178, see also **Code/source code**
    Arduino program structure: 56-58
    Basic code elements: 52-55
    Programming steps: 47-50
    Troubleshooting programs: 59-61, 155-159, 160-166

Programming board: see **FTDI board/ FTDI breakout board**

Programming Language: 46, 47, 74, 172, 173, **176**, 177, 178

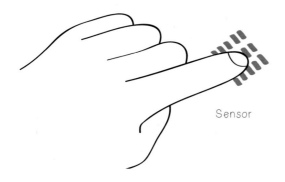

Sensor

Read (in programming, from an input device): 94, 95, 97, 99, 117, 137, 138, 148, 174, 175, 176, 177, 178, see also *digitalRead* and *analogRead*

Resistive sensor: see **Sensor, resistive**

Return (a value from a procedure): 95, 98, 116, 117, 176, **177**

Run (a program): 46, 47, 50, 51, 53, 56, 58, 87, 97, 100, 114, 132, 138, 147, 166, 174, 175,176, **177**, see also **Execute**

Running stitch: 13-14, 72, 93, **177**

Switch (circuit component): 17, **20**, 23, 35, 36, 37, 39, 43, 74, 92, 96, 161, 174, 176, **177**

Syntax: 48, **177**

Syntax error: 57, 173, **177**, see also **Compile error**

Sensor: 1, 18, 64, 74, 92-93, 94, 95, 96, 97, 98, 99, 100, 108, 110, 111, 112, 116, 117, 118, 121,126, 127, 136, 138, 146, 147, 148, 161, 162, 163, 166, 172, 174, 175, 176, **177**, see also **Input (device)**
    Capacitive: **110-111**, **118**, 123, 125, 126, 134-135, 161, **172**
    Resistive: **92-93**, 102-103, 110, **176**
    Troubleshooting sensors: 102-103, 123, 125, 134-135

Serial port: 45, 49, 60, 96, 103, 125, 147, 173, **177**

Serial.begin: 99, 114, 173

Serial.print: 136, 137, 146, 147, 177

Serial.println: 99, 114, 117, 131, 136, 138, 146, 147, 166, 177

setup/setup section (in Arduino programs): **56-57**, 58, 61, 80, 84, 91, 94, 96, 98, 99, 103, 114, 117, 121, 124, 127, 130, 141, 144, 161

Shock/electrocute: 20, 177

Short circuit/short: 9, 17, **19-20**, 33, 35, 60, 69, 71, 78, 79, 90, 102, 111, 123, 134, 151, **152-153**, **177**

Software: v, 2, 44, 45, 46, 47, 49, 54, 57, 58, 60, 73, 83, 94, 114, 116, 127, 136, 147, 148, 155, 156, 157, 158, 160, 173, 175

Source code: see **Code**

Speaker: see **LilyPad Speaker**

Statement/simple statement: 53, **54-55**, 57, 73, 74, 75, 78, 84, 94, 95, 99, 100, 117, 118, 120, 121, 127, 131, 135, 136, 137, 139, 140, 141, 142, 143, 161, 163, 164, 166, 172, 173, 174, 175, **177**, 178

Tab: **8**, 9, 10, 11, 13, 14, 15, 16, 19, 22, **28**, 31, 32, 33, 36, 37, 39, **43**, 66, 68, 69, 78, 81, 92, 94, 102, 111, 115, 116, 119, 137, 151, 152, 153, 154, 173, 176, **177**

Template: 8, 9, 10, 29, 32, 67, 70, 76, 93

Threshold: **100**, 103, 125, 163, **177**

Tilt switch: 23

tone: **82-83**, 84, 85, 120, 121, 124, 125, 130, 166

Trace/traces: **9**, 10, 13, 15, 16, 20, 22, 23, 28, 31, 32, 68, 69, 72, 78, 81, 93, 110, 111, 112, 126, 152, 153, 154, **178**

Troubleshooting solutions: **151-166**
    Code problems, general: 2, 78, 80, 91, 103, 124-125, 135, 146, 155
    Code problems, compile errors: **59**, 80, 91, 94, 103, 124, 135, 146, **155-159**
    Code problems, logical errors: **61**, 80, 91, 103, 124-125, 135, 146, **160-166**
    Electrical problems: 2, 17, 35, 78, 79, 90, 102, 123, 134, **151-154**, 160
    Sensor problems: 102-103, 123, 125, 134-135
    Software problems, upload errors: 45, **60**

Type/data type: 53, 54, 129, 139, 141, 172, 173, 175, **178**

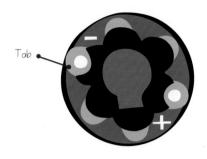

Tab

Upload: 46, 47, **49**, 50, 53, 54, 60, 61, 80, 91, 103, 124, 135, 146, 155, 163, 164, 175, 177, **178**

Upload errors: **60**, 61, 146, 155, 160, 175

USB cable/Mini-USB cable: 42, 44, 64, 73, 84, 96, 108, 114, **169**

Variable: 52, **53-54**, 55, 56, 57, 58, 61, 73, 80, 83, 84, 86, 91, 94, 97, 99, 103, 116, 117, 120, 121, 124, 125, 127, 129, 130, 131, 135, 139, 140, 141, 142, 143, 146, 155, 158, 160, 161, 162, 163, 165, 173, 174, 175, 177, **178**

Variable declaration: **53**, 56, 57, 140, **178**

Variable initialization: 53, 91, 124, 125, 135, 146, 155, 160, 161, 162, 165, 178

Voltage: 18, **19**, 74, 95, 98, 174, 176, **178**

Voltage rating (of a battery): 18, 19, 174

3 Volt battery

Washing instructions: 8, 21, 38, 43, 105, 149

Watt-hours: 18, 19, 174

while**/while loop**:142-144, 146, 163, 164, 165, 173, 177

Write (in programming, to an output device): 74, 176, 178, see also digitalWrite